José Miguel García, Félix Clemente García, José Antonio Reglero Ruiz,
Saúl Vallejos, Miriam Trigo-López
Smart Polymers

Also of Interest

Superabsorbent Polymers.
Chemical Design, Processing and Applications
Edited by Sandra Van Vlierberghe, Arn Mignon, 2021
ISBN 978-1-5015-1910-9, e-ISBN (PDF) 978-1-5015-1911-6,
e-ISBN (EPUB) 978-1-5015-1171-4

Sustainability of Polymeric Materials
Edited by: Valentina Marturano, Veronica Ambrogi, Pierfrancesco
Cerruti, 2020
ISBN 978-3-11-059093-7, e-ISBN 978-3-11-059058-6,
e-ISBN (EPUB) 978-3-11-059069-2

Processing of Polymers
Chris Defonseka, 2020
ISBN 978-3-11-065611-4, e-ISBN (PDF) 978-3-11-065615-2,
e-ISBN (EPUB) 978-3-11-065642-8

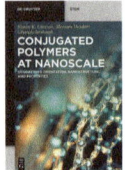

Conjugated Polymers at Nanoscale.
Engineering Orientation, Nanostructure, and Properties
Karen K. Gleason, Meysam Heydari Gharahcheshmeh, 2021
ISBN 978-1-5015-2460-8, e-ISBN 978-1-5015-2461-5,
e-ISBN (EPUB) 978-1-5015-1617-7

Multicomponent Polymers.
Principles, Structures and Properties
Guojian Wang, Junjie Yuan, 2020
ISBN 978-3-11-059632-8, e-ISBN 978-3-11-059633-5,
e-ISBN (EPUB) 978-3-11-059441-6

Polymer Synthesis.
Modern Methods and Technologies
Guojian Wang, Junjie Yuan, 2020
ISBN 978-3-11-059634-2, e-ISBN 978-3-11-059709-7,
e-ISBN (EPUB) 978-3-11-059714-1

José Miguel García, Félix Clemente García,
José Antonio Reglero Ruiz, Saúl Vallejos,
Miriam Trigo-López

Smart Polymers

Principles and Applications

DE GRUYTER

Authors

Prof. José Miguel García
jmiguel@ubu.es
Universidad de Burgos
Facultad de Ciencias
Departamento de Química
Plaza de Misael Bañuelos s/n
09001 Burgos
Spain

Prof. Félix Clemente García
fegarcia@ubu.es
Universidad de Burgos
Facultad de Ciencias
Departamento de Química
Plaza de Misael Bañuelos s/n
09001 Burgos
Spain

Dr. José Antonio Reglero Ruiz
jareglero@ubu.es

Dr. Saúl Vallejos
svallejos@ubu.es
Universidad de Burgos
Facultad de Ciencias
Departamento de Química
Plaza de Misael Bañuelos s/n
09001 Burgos
Spain

Dr. Miriam Trigo-López
mtrigo@ubu.es
Universidad de Burgos
Facultad de Ciencias
Departamento de Química
Plaza de Misael Bañuelos s/n
09001 Burgos
Spain

ISBN 978-1-5015-2240-6
e-ISBN (PDF) 978-1-5015-2246-8
e-ISBN (EPUB) 978-1-5015-1545-3

Library of Congress Control Number: 2021951144

Bibliographic information published by the Deutsche Nationalbibliothek
The Deutsche Nationalbibliothek lists this publication in the Deutsche Nationalbibliografie;
detailed bibliographic data are available on the Internet at http://dnb.dnb.de.

© 2022 Walter de Gruyter GmbH, Berlin/Boston
Cover image: peepo/iStock/Getty Images Plus
Typesetting: Integra Software Services Pvt. Ltd.
Printing and binding: CPI books GmbH, Leck

www.degruyter.com

Contents

1 General introduction and preface

Nature is full of examples of stimuli-responsive (or smart) materials. Leaves of the *Dionaea muscipula* (venus flytrap) close capturing insects, leaflets of *Codariocalyx motorius* (telegraph plant) and *Helianthus annuus* (sunflower) rotate under exposure to sunlight, the former at a speed that can be visually perceived, leaves of *Mimosa pudica* (sensitive plant) can collapse when shaken or touched and *Chamaeleonidae* (chameleon) or *Octopoda* (octopus) change their colour depending on the environmental condition or situation. These natural phenomena have attracted the interest of researchers for a long time, and many different efforts have been carried out to mimic this behaviour using synthetic materials. In this sense, the synthesis, characterization and applicability of stimuli-responsive polymers have become one of the most important research lines of the polymer science.

Smart polymers present the ability to respond to different stimuli by changing their physicochemical properties. The type of response varies from dimensional variation (shape-memory polymers), to changes in colour, electrical conductivity, luminescence and many others. Concerning the stimuli, we can have, for example, temperature variations (thermo-responsive), pH (pH-responsive), magnetic and electric fields (magneto-responsive or electrical-responsive polymers), humidity or light (light-responsive polymers). Furthermore, when the external stimulus is the chemical or physical interaction between the polymer and an external substance itself, smart polymers can be employed as sensory materials due to their ability to detect and quantify the target substance by analysing their responsive properties. Here, the term "smart" indicates that the polymer exhibits a specific response for a given stimulus; thus, the control, triggering and analysis of the stimuli-response relation in smart polymers is the key factor in which researchers put their efforts.

This book describes the fundamentals and main applications of smart polymers. Chapter 2 will be devoted to resume the fundamentals of smart polymers, listing the main families of responsive polymers in terms of external stimuli, such as in pH, temperature, light and mechanical stimuli. In Chapter 3, we will describe the use of smart polymers as key components of sensory systems. Due to their intrinsic nature, stimuli-responsive polymers can be designed to respond to a wide variety of different stimuli, resulting in changes in shape, solubility, surface properties, colour or fluorescence, including also a section dedicated to the immobilization of biomolecules, which is especially interesting for the recent developments in the recognition and immobilization of several proteins which are directly related to several diseases (e.g. the immobilization of the human angiotensin-converting enzyme 2 allows the recognition and detection of the SARS-Cov-2 pathogen). Additionally, the easy processability of polymers allows them to be manufactured into different forms such as films, beads, coatings and fibres, making them promising materials for sensors fabrication. For these reasons, the use of smart polymers as sensory materials

https://doi.org/10.1515/9781501522468-001

was one of the earliest developments in this research field. However, the exponential interest in biomedical and biological-related applications over the last decade focus the attention of Chapter 4, becoming an essential part of the book. With the obtention of new biocompatible and biodegradable smart polymers, and the designing and fabrication of hydrogels, that can be easily implemented in biological media with minimal risks, the use of smart polymers in these applications have significantly increased. We will analyse four different research lines in which these materials play a key role: drug delivery, tissue engineering and precision medicine or cell therapy. The use of smart polymers in precise drug delivery implies the use of pH-responsive polymers that can retain a specific drug when entering the stomach, where it could irritate or inflame the stomach lining, but then rapidly release it when it reaches the intestines where the pH rises to physiologic pH levels. Tissue engineering employs regenerative polymers that present similar properties to human skin or tissues, taking advantage of their biocompatibility and/or biodegradability. Finally, precision medicine uses pH-responsive polymers that can interact directly with a specific cell or a group of cells to treat directly diseases such as Alzheimer or brain tumours, due to the different pH that these cells present with respect to healthy ones. Our final chapter will include some perspectives and reflections about the evolution of smart polymers, that will have undoubtedly to turn their main efforts to study their role in the biomedical and biological fields, broadening their applicability and leading the materials science in the future years, driven by the aim to provide solutions to societal challenges. In this sense, polymer science is continuously evolving, and there are still numerous polymer structures that remain not investigated, methods to be developed for their synthesis and new properties to be discovered.

In short, we present here a critical review of the state of the art of this outstanding research field, including both the chemico-physical fundamentals and main applications of smart polymers. However, this text comes from our experience as researchers and authors in sensory polymers, so our personal view has been relevant for writing the book, and this must be considered by the reader. In this way, the classification in responsive and sensory polymers, commented in the next chapter, is affected by the mentioned research experience. We believe that this book can be used as a teaching manual in an advanced polymer science course, but also it can be very useful to the polymer science researchers, as a reference guide for novel researchers that takes their firsts steps in the polymer science, and to the experienced scientist working in a specific polymer science area.

2 Fundamentals of smart polymers

2.1 Introduction

The term polymer, and more specifically polymeric, was coined in 1832 by Berzelius [1]. Parallel to the development of chemistry as a science, and in complete analogy to its evolution, the industrial importance of polymer science and technology began long before establishing the scientific foundations of science itself. Although cellulose was isolated and chemically modified throughout the nineteenth century as a substitute for silk and ivory, the real breakthrough in industrial production of modified natural polymers came in 1839 with the vulcanization of natural rubber by Goodyear, and concerning fully synthetic polymers in 1908, by Baekeland, with the development of phenolic resins known as Bakelites.

However, despite the commercial success of polymers or macromolecules, at the beginning of the 1920s, their structure was completely unknown since the colloidal theory was in force, which stated that polymers were simple associations of molecules of relatively small molecular mass. Thus, in 1920, Staudinger, professor of organic chemistry at the ETH in Zurich, published an article entitled "Über Polymerisation" [2], which describes various polymer formation reactions called polymerization, in which discrete molecules react with each other, giving rise to structural units that are repeated by formation of conventional covalent bonds. This radically new concept of polymers, referred to by Staudinger as "Makromoleküle" in 1922 [3], coining the term macromolecules, covered both synthetic and natural or modified polymers, reinforcing his proposition of polymers as long polymer chains, thus laying the foundations of polymerization and polymer science and technology.

The development and description of the structure of macromolecules paved the way to the precisely designed architecture of synthetic polymers to meet the technological needs of today's highly technical society. Thus, the buildout of conventional random chain or step-growth polymerization and copolymerization techniques based on radically initiated polymerization or on well-known organic reactions afforded an impressive number of structures leading to polymers designed to have target properties. However, recent polymerization techniques allowed for the easy preparation of precise structures, in terms of blocks and chain size and distribution that are basic for fine-tuning the properties of responsive polymers. Among these techniques are the evolution of anionic polymerization, that is, controlled/living radical polymerization (e.g. atom transfer radical polymerization, reversible addition-fragmentation chain transfer polymerization and nitroxide-mediated polymerization [4], and chain growth polycondensation [5]. Also, highly efficient techniques for modifying polymers, in terms of yield, lack of side reactions, smooth conditions and tolerance to functional groups, such as "click chemistry", have fostered the

https://doi.org/10.1515/9781501522468-002

evolution of smart polymers from the viewpoint of design of the polymer and the response mechanism [6].

Thus, the science and technology of polymers is directed today towards obtaining and studying special polymers, prepared directly by synthesizing new monomers or through chemical and physical modifications of pre-existing polymers, and it has become a cutting-edge science, eminently interdisciplinary, which is on the frontiers of chemistry, physics, engineering and biology, and which also requires knowledge about synthesis, characterization, structure, processing, properties and behaviour of materials. The main objectives of this branch of scientific research focus on the preparation of materials with high modulus, high thermal and oxidative resistance, non-flammable, electroactive, photosensitive, biopolymers, polymers with non-linear optical properties, nanomaterials, multicomponent systems with special properties, selective materials for separation or analysis techniques, for medical applications, with biodegradable structures, support for heterogeneous catalysis or for the automatic synthesis of proteins or nucleic acids and so on.

In this way, polymers are currently part of a wide range of structural, functional and special application materials, which find application in the construction, aeronautical, automotive, container and packaging industries, electronics, in medical applications and so on, acquiring great importance in the economy and social welfare. The diversity of applications that polymeric materials find is due to the variety of physical and chemical properties that they can present, which are intrinsically related to their structure, which derives from the nature of the monomer and the bonds they form throughout the polymer chain, in addition to the length and functionality of the side groups that they may incorporate. This extraordinary development has notably boosted research in this field, one of the most active from a scientific and technological point of view today. The general advances in science and technology made in recent decades are mainly due to the rapid evolution of polymer science and technology, which has become a key instrument in humans' development, safety, and quality of life.

The inherent macromolecular structure of polymers turns these materials sensitive to each macromolecular chain's microenvironment, regardless of the state, solid or solution. Thus, several variables, such as temperature, humidity and mechanical stress, change the way each structural unit of each chain interacts intra- and intermolecularly with others and with solvents, absorbed species and so on. And this leads to changes that can be used to sense this microenvironment. The question is if we have a technique with the sensitivity to correlate these changes with a useful output or signal, thus entering into the field of smart polymers. This challenge is faced by chemically designing sensory monomers to get sensory polymers with easily measurable responses to specific targets, where these targets can be a physical stimulus (temperature, light, electrical and magnetic, mechanical, etc.), a chemical stimulus (pH, solvent, redox reaction, chemical species, etc.) or a biomolecule in biological media (enzymes, proteins, glucose, etc.).

The ability to respond to changes in environmental conditions is the key factor of the functionality of biomacromolecules, such as proteins, DNA and RNA, and the milestone of life itself, intimately related to sensing mechanisms of complex nature. And the goal in the design of advanced intelligent materials is intimately related to mimicking nature in selective and sensitive recognition, structural accommodation and response [7].

This book will show the potential of smart- or stimuli-responsive polymers as sensory materials concerning their ability to respond to different physical or chemical stimuli, such as temperature, electromagnetic pulses, pH, chemical species or biological molecules, changing different physicochemical properties (e.g. solubility, colour, fluorescence, electric conductivity or shape) [8]. The smart behaviour of these materials is boosted by the shapes that can be obtained by simple transformation of polymers, such as coatings, films, fibres or wires, coupled with the easy tuning of key properties, such as their hydrophilicity, porosity and mechanical properties. Also, the relevance of smart polymeric materials in biological and biomedical applications will be reported, for example, in the diagnosis of numerous diseases, selective release of drugs at cellular level, tissue engineering and regenerative medicine, biosensors for detection and immobilization of biomolecules and cell therapy and precision medicine.

There are multiple ways of classifying this kind of polymers and, in this sense, the type of response has been used in this manuscript for the classification into polymers that respond to the stimulus with an action, for example, drug delivery polymers, from which the generic term responsive polymers is used, or with an alert consisting of useful analytical information, termed herein sensory polymers. Accordingly, the same stimulus, both physical and chemical, can affect a sensory polymer to give a macroscopic processable signal or by a responsive polymer to perform a specific action.

2.2 Response to physical stimuli

2.2.1 Temperature as stimulus

Temperature can be easily monitored and controlled. For this reason, it has been widely used as a stimulus in responsive polymers. Temperature affects a polymer solution in terms of the interaction of the polymer with the solvent. Thus, when phase changes are observed at a given temperature, we have a thermo-responsive polymer system comprised of a pair polymer/solvent. Solution systems, a single-phase, therefore, can split into two phases by heating or cooling, thus exhibiting an upper critical solution temperature (UCST) or a lower critical solution temperature (LCST), respectively [9].

LCST-exhibiting polymers have been broadly studied, and the structure of the polymers modified to tune the "cloud point" at which the transition between a transparent solution to opaque is observed. The one-phase solution comes from strong interactions of the structural units of the polymer with the solvent, usually hydrogen bonding with water. Upon heating, the hydrogen bonds are weakened and the inter- and intra-chain interaction reinforced, finally leading to the aggregation of the polymer chains due to the dominance of hydrophobic interaction above the LCST.

On the other hand, UCST exhibiting polymers have been much less studied. The transition from the globule state to the open coil state is enthalpy driven, the entropic (hydrophobic effect) is less dominant, and it can depend on hydrogen bonds or Coulomb interactions. In addition, there may be direct inter- and intra-chain interactions or bridged by water or ions [10].

2.2.2 Light as stimulus

Light is an interesting stimulus that can be controlled and measured from multiple aspects, such as wavelength, intensity, area of exposure or a combination thereof. Light-responsive polymers are mainly based in two approaches: photochromic and photocleavable groups [11].

Polymers having photochromic groups undergo isomeric changes upon irradiation at a given wavelength, and sometimes these changes are reversible by irradiating at another wavelength of at a given temperature. These changes are usually *cis–trans* isomerization, intramolecular hydrogen or groups transfers or pericyclic reactions. These changes modify the chemical structure with other physical (electrical, optical, mechanical, etc.) and chemical properties, for instance, with different chromatic of fluorogenic properties or other physical characteristics, such as refractive index and dipole moments and geometrical structure. The proper device can record the response, or even visually, to get information about the environment [12–14].

Polymers with photocleavable groups in their structure usually undergo chemical changes at a given wavelength, usually irreversible, giving rise to other functional groups that provide the materials with different physical and chemical characteristics [15].

2.2.3 Electrical stimuli

Electrical stimulus may induce changes in the shape of electro-responsive polymers, usually conductive or ionic polymers. This stimulus can be easily controlled through the current magnitude and the electrical pulse to transduce it into mechanical work for actuators [16, 17].

2.2.4 Mechanical stress as stimulus

High mechanical stress of macromolecules gives rise to a reduction of the average molecular mass, and is usually observed in the transformation of polymers, usually termed as mechanical degradation. This degradation comes from the unspecific homolytic bond cleavage, followed by unspecific side reactions or chain rearrangements. However, the term mechano-responsive polymer comes from the fact that the mechanically driven bond cleavage or rearrangement can be targeted to mechanically labile groups, called mechanophores, precisely designed to impart smart capabilities to the materials. The cleavage or rearrangement can be observed in solution, in flow field and under ultrasounds, and in the solid state usually in compression or elongation mechanical forces [18, 19].

2.3 Response to chemical stimuli

2.3.1 Chemical species as stimulus

The work carried out by Pedersen, Cram and Lehn in the 1960s lead to the development of supramolecular chemistry, paving the way to the dawn of a new research field called chemical sensors or chemosensors, where low molecular mass molecules, termed hosts or receptors, selectively interact with target chemical species, acting as a stimulus, also called guest, with an action or a concomitant measurable change in a macroscopic property, or transduction. After the initial steps of host–guest chemical sensing by low molecular mass chemosensors, the selective interaction with synthetic species was extended to polymers with receptor motifs in their main or lateral structure, giving rise to a sensory polymer when a change in a measurable physicochemical property is obtained, such as colour, fluorescence or resistivity, or to a responsive polymer, when an action is obtained, such as a change in shape or solubility.

Polymers, as macromolecules, have important advantages compared to low-mass chemosensors, such as the lack or migration of the sensory motifs, and the possibility to prepare the sensory materials in several shapes, such as films, coatings, beads and fibres. Also, as biomacromolecules, they may exhibit the allosteric effect and other collected properties and signal-amplification characteristics.

A wide assortment of target chemical species can be identified or trigger the smart polymers' action. Among the chemical species of interest to identify, it is worth mentioning heavy metal cations and anions, acidity, organic volatile compounds and gases, chemical warfare agents, and biomolecules, such as proteins, disease markers and glucose. Also, these species, and specifically biomolecules, can trigger the response of responsive polymers, for instance, shape and solubility changes leading to drug delivery [8, 20, 21].

2.3.2 pH as stimulus

Polymers respond to the medium's acidity when they have pendant or main chain groups that can accept or donate protons. Accordingly, variations in pH in the microenvironment of these sensory polymers induce changes in the main or lateral polymer chains, in terms of inducing a partial polyanion or polycation structure. This strongly modifies the solvent-polymer and inter and intrachain interactions, giving rise to the collapse or expansion of the macromolecules due to coulombic interactions, affecting solubility in solution and shape in the gel state. Also, the overall electronic structure of the system is affected, changing the highest occupied molecular orbital–lowest unoccupied molecular orbital and, hence, the spectroscopic behaviour, specifically fluorescence and colour, especially in polymers having groups with charge-transfer complexes. Gels and solid-state sensory materials can also sense vapours, for instance, detecting changes in the acidity of the atmosphere surrounding the smart materials [22–24].

2.3.3 Solvent as stimulus

Polymers can interact with solvents in solution and in the solid-swelled state, where the swelling and de-swelling processes depend on the solvent or the composition of a mixture of solvents. This interaction may change the shape or microstructure to achieve functionalities in finished goods or even to sense by a shape change, chromogenic behaviour or other easily macroscopic property, the solvent, solvent mixture or even water content of organic solvents. Obviously, these changes can also be exploited to sense vapours, for example, water or methanol, by a colour change (vapochromism) due to structural change of solid-state polymers coming from the selective interaction of macromolecules with solvent in the gas phase [25, 26].

2.3.4 Redox as stimulus

Polymers having chemical moieties sensitive to oxidations and reductions undergo changes in oxidations states, thus modifying the electronic structure of these groups and, accordingly, that of all the polymeric chains. These changes are responsible for variations in the hydrophilic/hydrophobic balance of the material with the concomitant swelling/de-swelling in the solution media, changes in colour or fluorescence, shape and so on [27–30].

2.3.5 Varied stimuli that trigger controlled depolymerization

Oligomers, dendrimers and macromolecules that undergo controlled depolymerization promoted by a stimulus-responsive trigger are known as self-immolative polymers, which can undergo head-to-tail depolymerization when triggering moieties, at the end of the polymeric chains, are activated under selective responsive triggers, such as light, pH, enzymes, redox and chemical species.

The applications of self-immolative polymers are related to the fact that the interaction of a single stimulus triggers the depolymerization of a whole polymer chain releasing multiple small chemical species, covalently bound to the chains, following an amplification pattern of the stimulus [6, 31].

2.4 Response to biological stimuli

The concept of using small to moderate environmental stimuli within biological media, such as chemical species, pH, light and temperature, to obtain a response in smart polymeric materials in an endeavour that mimics the smartness behaviour of biological systems in biological media. Thus, the field faces the fascinating interface between chemistry and biology to open a broad set of possibilities, applications and tools in biosensing, tissue engineering, precision medicine and cell therapy, immobilization of biomolecules and so on.

Obviously, biological stimuli are physical or chemical stimuli. However, they have been considered in a specific section centred in applications due to the impact of the smart polymers in biological applications of the present and forthcoming derived applications, in terms of benefits for society and increase in the welfare of humankind. Thus, it is a highly evolving area full of opportunities related to research and innovation.

2.4.1 Polymer-based biosensors

From the early discovery of synthetic macromolecules, polymers have been widely used in biochemical science, where the term intelligent, or smart polymeric materials, points out to a broad set of synthetic, natural or chemically modified natural polymers with unique properties for biological applications. The inertness, versatility related to the chemical constitution and design, and their macromolecular nature of polymers, like relevant biopolymers, such as proteins or DNA/RNA, make them ideal materials of advanced biological applications where chemistry meets biology.

The field of biosensing by using polymer-based biosensors is fostered by the advantage of using polymers in terms of specificity, versatility, easy design, low cost and so on. Among these advanced applications of stimuli-responsive polymers, the development of biosensors implies mimicking nature in recognition processes where the selective interaction of parts of a polymeric structure acting as host or receptor, motifs and a guest or target species, gives rise to a change in a property than can be easily detected, registered and processed to get information of the presence and concentration of the target. Of special significance in diagnosis is the supporting and entrapping or immobilized antibodies, antibody fragments or enzymes and DNA. Target biomolecules are every biologically significant biomolecules, such as glucose [32–34].

Thus, polymers bearing specific sensitive moieties to enzymes, or enzyme-substrate, in the main or lateral chains provide these polymers with smart characteristics, for instance, for selectivity cleaving a particular enzyme. In a similar fashion, antigen/antibody recognition can be exploited to prepare polymers with antigen/antibody and peptide substrate sensing capabilities. Also, other relevant biomolecules, such as glucose, are target objectives in the design of sensory polymers having selective glucose receptor motifs in their structure or chemically anchored or entrapped glucose oxidase [33, 35].

2.4.2 Drug delivery

Systems devoted to drug delivery involved technologies for the controlled release of therapeutic agents following a pattern, initially at an established rate, and responding to stimuli arising from the illness's evolution in a regulated drug release [36–38].

Initially, polymer carriers were used as vehicles to promote the time-controlled release of therapeutic agents in both pulsatile and implanted delivery systems. These conventional drug delivery systems have highly improved the treatment of diseases, paving the way to the dawn of smart delivery systems, which are facing the need for specific targeting while responding precisely to physiological environments, and at the same time, addressing challenges as biocompatibility and intracellular transport. Polymers offer an easy control of their functionality and physicochemical features because they can be easily designed and prepared with different architectures (linear, cross-linked, hyperbranched, etc.) and functional groups to active biochemical characteristics, such as biocompatibility and hydrophilicity/hydrophobicity balance, and at the same time, they are stable, versatile and not expensive [39].

2.4.3 Tissue engineering

In vitro or *in vivo* tissue engineering is devoted to maintenance, restoration, improving tissue functions or even substituting complete organs. From a preparation viewpoint of synthetic materials for this purpose, the challenge is to have materials with similar properties to human tissues, playing biocompatible and biodegradable polymeric hydrogels a key role as scaffolds in this endeavour [40, 41].

2.4.4 Precision medicine and cell therapy

Precision medicine and cell therapy are devoted mainly to diagnosing relevant diseases and the treatment of cancer. They have the potential of allowing the early detection and prevention of diseases, and the guidance and control of the release of bioactive chemical species to the therapeutic target organ at the correct timing and concentration. In this sense, smart polymers, as intelligent soft materials that respond to physical or chemical environmental conditions, play a relevant role in this field. Thus, thermo-responsive, shape-memory and self-healing polymers are used as cell/drug/protein carriers. They can be applied in precision medicine, bioprinting and minimally invasive surgery [42].

2.5 Multiple stimuli-responsive polymers

Dual and multiple stimuli-responsible polymers are prepared with a set of monomers bearing in their chemical structure groups that separately impart to the polymers prepared with them a smart behaviour to a single stimulus. Thus, the combination of these monomers can provide smart materials with response to thermal and light changes; or to thermal and pH variations; or to thermal, pH and light changes or with thermal, redox and light response [43, 44].

3 Sensory polymers

3.1 Introduction

Sensory polymers have the ability to respond to different stimuli, through a specific mechanism, to generate a certain response [8]. Three different classifications of sensory polymers are reported, depending on the stimuli, the mechanism and the response, as graphically depicted in Figure 3.1.

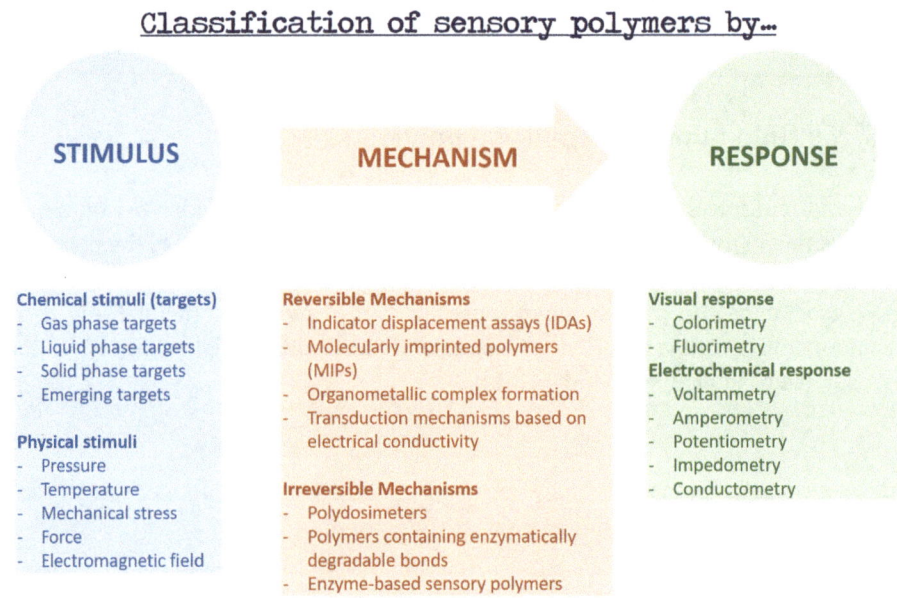

Figure 3.1: Graphical abstract of the different types of sensory polymers, classified by stimulus, mechanism and response.

Unlike other smart polymers (SPs) described in this book, such as drug delivery systems (**Section 4.2**), which respond to the stimulus with an action, the response generated by the sensory polymers is an alert. Ultimately, they transform certain chemical or physical information into useful analytical information [8, 11].

In the same way, it is also important to clarify that responsive polymers and sensory polymers usually have very similar chemical composition, but the main difference relies on the application for which they are designed: sensitive polymers respond with an action, and sensory polymers respond with an alert. Thus, a thermo-responsive polymer can shrink and release a drug with increasing temperature, while a thermochromic polymer responds to the same stimulus with a colour change [45].

https://doi.org/10.1515/9781501522468-003

Depending on their chemical structure, sensory polymers can be formed by a single monomer (sensory homopolymers) or by several monomers (sensory copolymers), taking multiple shapes, such as linear, branched or cross-linked polymers.

Sensory copolymers usually combine functional monomers or groups designed to interact with the stimulus, called sensory monomers or sensory groups. On the other hand, these monomers can be copolymerized with non-sensory monomers or structural monomers, which do not interact directly with the stimulus but allow to achieve specific properties in the final copolymer such as electrical conductivity, hydrophilicity, hydrophobicity and affinity for a substrate.

Sensory polymers can be transformed easily into different shapes, such as coatings, films, fibres or wires, tuning at the same time their hydrophilicity, broadening the applicability of SPs in many different sensory devices, from the detection of harmful substances (volatile organic compounds [46] or heavy metal cations [47]), to the most recent and outstanding applications in the detection of molecules, virus, bacteria and so on in biological media.

To conclude, the last section of this chapter will be devoted to resume the main sensory polymers and applications in which this kind of SPs are employed, covering both classical and recent developments, and focusing our attention on specific and key research works in the field of polymeric sensors, from colorimetric and conducting polymers for gas detection or Hg(II) quantification in aqueous media, to the recent advancements in the use of SPs in textile applications or as biosensors.

3.2 Classification of sensory polymers by stimulus

Two large groups of stimuli can be distinguished: chemical stimuli (also called species of interest or targets) and physical stimuli (such as temperature and pressure).

3.2.1 Chemical stimuli: targets

Concerning chemical stimuli, targets are generally the species of interest to be identified and/or quantified. These polymers arise from a problem related to these chemical species (targets), and therefore, they are born as a solution that is usually quick, simple and direct [20]. Thus, there is a broad list of targets, ranging from everyday problems such as determining the pH of a swimming pool to advanced solutions related to diagnosing and controlling diseases [48, 49].

Targets can be in the gas phase, in solution or they can even be solids. Here is where sensory polymers outstand over non-polymeric sensors, being adaptable not only to the specific target, but also to the environment where they are found. This adaptation depends not on the sensor units but on the rest of the polymer chain and how the polymer is transformed.

Undoubtedly, the most explored scenario is the detection of targets in solution, such as enzymes in the biological media or pollutants in drinking water. In this case, the detection phenomenon can be approached in different ways. When a manageable solution is required, the best choice is to use dense membranes or films that combine a sensory monomer to detect the target species and generate a response with other hydrophilic monomers that give rise to a material with gel behaviour to be used in aqueous media. Also, using coatings based on sensory polymeric electrodes is a very recurrent and relatively easy way to work, producing excellent results with simple procedures.

The detection in the gas phase does not depend so much on the hydrophilicity of the sensory polymer since the interaction occurs between a solid (sensor polymer) and a gaseous target. Therefore, this type of solution can also be approached with sensory polymers as films [50], although it can also be carried out by coatings based on linear or branched sensory polymers containing specific receptors, and monomers specifically selected to provide affinity with the substrate. In this way, sensory systems based on paper or textile fibres can be obtained, giving outstanding results in the detection of targets in the gas phase [51].

Finally, although less explored, sensory polymers can also be designed to detect solid targets so that molecular recognition occurs between two solids. The most relevant applications that have been described for this type of sensor polymer range from the detection of traces of heavy metals in the hands of an operator [52], to the manufacture of smart labels that comply with the food-contact regulations for the quality control of food by the final consumer [53].

The mere existence of sensory polymers is intimately related to the targets. In this section of the book, we briefly discuss the most common targets and the increasing pollutants that have emerged in recent years, which pose a challenge to society and the scientific community.

One of the pollutants family that worries society is the persistent organic pollutants (POPs), which were intentionally and unintentionally produced and are now ubiquitous contaminants in soil, water and air. Due to their persistence and lipophilicity, they bioaccumulate and are very hard to degrade. Pesticides (dichlorodiphenyltrichloroethane, DDT), polytetrafluoroethylene (PTFE) coatings (they have a practical use in frying pans or textiles as well as perfluorooctanoic acid or perfluorooctane sulfonate), sophisticated fire-fighting foams used by fire brigades (perfluoroalkyl substances, PFAS) and some building materials (polychlorinated biphenyls or hexabromocyclododecane) are only a few examples of new pollutants, of which sensory polymers science will have to tackle in the following years. On the whole, these substances can be divided into four families of POPs to give a complete overview of all existing groups: (1) halogenated aromatic POPs, (2) halogenated non-aliphatic POPs, (3) aromatic POPs and (4) aliphatic POPs.

Additionally, the Joint Research Centre from the European Commission report shortlists of potential water pollutants to complete a first Watch List which will provide high-quality information on the concentrations of emerging or little-known pollutants across the EU [54]. The report proposes some prescription drugs (for instance, diclofenac, 17-beta-estradiol, 17-alpha-ethinylestradiol, erythromycin, clarithromycin, azithromycin), some pesticides (methiocarb, tri-allate, imidacloprid (thiacloprid, thiamethoxam, clothianidin, acetamiprid)) and some personal care products (2-ethylhexyl 4-methoxcinnamate, 6-ditert-butyl-4-methylphenol). This tendency has also been observed by other countries such as India, which has discovered that 57% of the studied contaminants were pesticides, 17% were pharmaceuticals, 15% were surfactants, 7% were personal care products and 5% were others.

3.2.2 Physical stimuli

Sensory polymers for detecting physical stimuli are much less abundant, and in many cases, it is not easy to differentiate between a stimuli-responsive polymer and a sensory polymer. In fact, this book mentions physical stimuli-responsive polymers in **Section 2.2**, and this section is only intended to give a few brushstrokes of the most relevant applications of this type of polymers.

For this kind of sensory polymers, the most common uses are as thermochromic polymers in devices to control room temperature or even body temperature [45, 55].

Additionally, related to advanced applications, physical stimulus as mechanical stress (mechanochemoelectric effect) can be detected using polymer-based films of three-layer polypyrrole (PPy) sensor. In this sense, Figure 3.2 shows an example of how mechanical deformation can produce measurable electronic currents. In terms of produced charge, these sensory polymers go beyond other ones as piezopolymers that convert pressure to voltage, but deep down, they are based on the same concepts. The mechanical stress increases ion concentration and mobile ions are expelled from the contracted layer on the opposing side. The generated potential difference between two layers can be detected by an open-circuit potential measurement, or by detecting a short-circuit current [56,57].

In the same way, equivalent sensory polymers can be designed for the detection of other physical stimuli as a force [58] or an electromagnetic field [59], and based on different polymers as poly(N-isopropylacrylamide), poly(methyl methacrylate) (PMMA) or poly(vinyl alcohol).

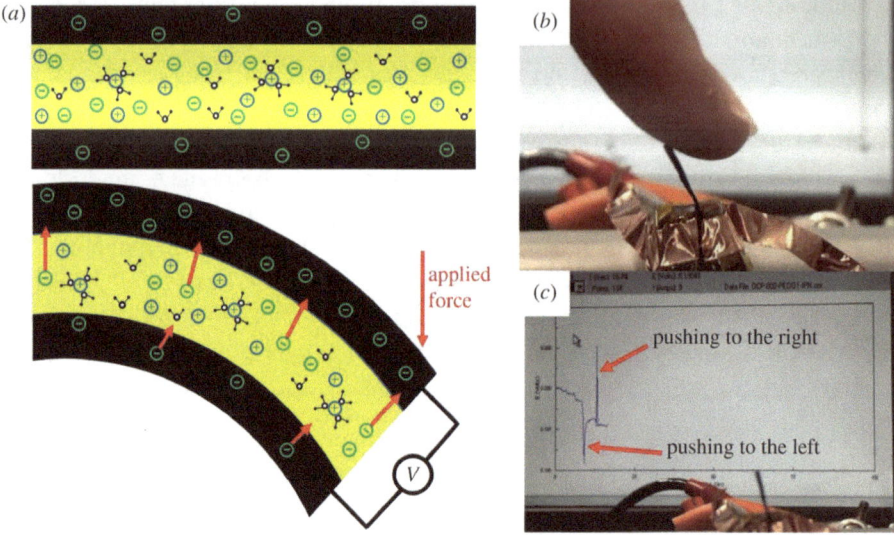

Figure 3.2: Mechanochemoelectric effect in a polypyrrole three layer sensor: (a) bending of the three layer sensor; (b) a three layer sensor is detected by a finger; (c) open circuit potential generated by the mechanical stimuli. Reproduced with permission of The Royal Society [57].

3.3 Classification of sensory polymers by mechanism

3.3.1 Reversible mechanism-based sensory polymers

3.3.1.1 Indicator displacement assays

Sensor polymers based on indicator displacement assays (IDAs) are always linked to a visual response, usually colorimetric or fluorometric. What is more, this type of sensory polymers is postulated as an advantageous alternative when the interaction between the target and the receptor units does not generate a coloured or fluorescent species. That is why an indicator species, a chromophore or a fluorophore, is included as an extra component in their formulations. This indicator species is recognized by the receptor units of the sensor polymer in the same way as the target, but through a weaker interaction. Otherwise, from an analytical point of view, if the interaction with the indicator is stronger than with the target, the indicator would be an interference and the sensor system would not be valid. Consequently, when the target reacts with the receptor units, it displaces the coloured or fluorescent indicator to the medium and produces a visual change (Figure 3.3) [60–62].

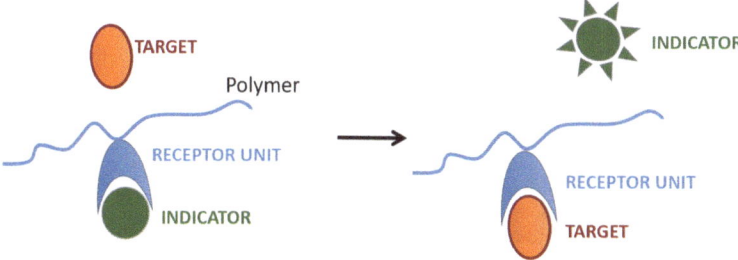

Figure 3.3: Schematic representation of an indicator displacement assay.

3.3.1.2 Molecularly imprinted polymers

Molecularly imprinted polymers (MIPs) are especially interesting in the field of sensory polymers. They are prepared by a process in which the sensory and cross-linking monomers are copolymerized in the presence of the target, which acts as a molecular template. The sensory monomers initially interact with the target, and after polymerization, their functional groups remain in the same position due to the highly cross-linked polymeric structure. Subsequent removal of the imprinting molecule reveals binding sites [63], complementary in size and shape to the target. In this way, a molecular memory is introduced into the polymer, which can now reattach the target with very high specificity [64–66].

This methodology is also used to prepare other kinds of SPs, with different objectives, as the immobilization of biomolecules (**Section 3.6**). Indeed, the use of MIPs as sensory materials emerged a few years ago, as was pointed out in 2007 by McCluskey *et al.* [67], and nowadays, it presents interesting application perspectives in many fields such as drug delivery, biotechnology, separation sciences and chemo/biosensors [68]. Among many other functional monomers, methacrylic acid, 4-vinylpyrrolidine, acrylamide or hydroxyethyl methacrylate are used to implement the binding interactions in the imprinted sites, taking into account that cross-linkers and solvents must be appropriately selected to confer the mechanical stability and processability required.

One of the key factors considering the importance of MIPs as sensors is their high selectivity and easy preparation, especially relevant in detecting emerging pollutants or new targets. Figure 3.4 shows the schematic fabrication process of MIPs as sensory polymers [69].

3.3.1.3 Organometallic complexes formation

These sensory polymers contain receptor units that interact with different metal ions. The formation of these organometallic complexes in the polymer chain is often favoured compared to the same non-polymeric sensory system, that is, with conventional probes [70]. The formed organometallic complexes have different

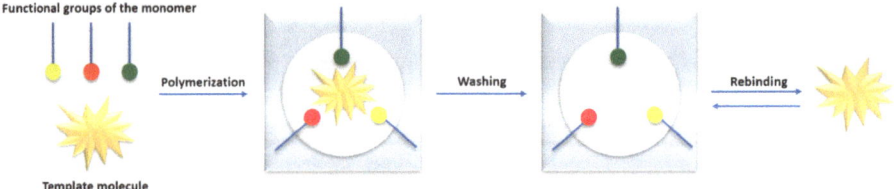

Figure 3.4: Schematic fabrication process of molecular imprinted polymers (MIPs). Reproduced with permission of MDPI [69].

characteristics derived from the ligands (or receptors), and in many cases colour or fluorescence changes occur, and this fact is the basis of these kinds of sensor polymers.

In this book, these types of sensory polymers have not been included as chemical dosimeters, since in general, the formation of organometallic complexes is reversible under the appropriate conditions.

3.3.1.4 Transduction mechanisms based on electrical conductivity

Sensory polymers based on this type of mechanism are conductive (or conjugated) polymers. Compared to discrete receptors in solution, that is, conventional probes, this type of polymers offer a much more sensitive collective response to the presence of a stimulus due to its transport properties, electrical conductivity or rate of energy migration. Especially with conductive polymers, it is important to differentiate between the type of response of a sensory polymer, and the type of mechanism by which this response occurs. That is why, in this book, conductive polymers are in this classification [71–74].

As mentioned, the response generated by a conductive polymer-based sensor can be analysed by voltammetry, amperometry, potentiometry, impedometry, conductometry and so on. This is intuitive, as it is easy to relate an electrical response to the intrinsic electrical conductivity of these sensory polymers. But these polymers can also generate visual responses (colorimetric and fluorometric), and also in these cases, the response is a direct consequence of the conductive nature of these polymers. Thus, this section intends to describe the mechanism by which this type of polymers mediates or transduces responses that indicate the presence of a stimulus.

Conductive polymers can be found both in their pristine (neutral) state and in their doped state. The adjective "doped" has been inherited from the most common semiconductor materials, which are certain metals such as germanium. However, it is important to clarify that when we speak about conducting polymers, the term "doped" refers to modifying the π-electronic system and not to atoms replacements. The p-doping implies an oxidation of the π-electronic system, and the n-doping a reduction. The doping process of a conductive polymer can be carried out chemically or electrochemically, and it is necessary to incorporate a counter ion to maintain the electroneutrality of the system.

The chemical structure of conductive polymers is responsible for their being considered wide band gap semiconductors. Conductive polymers such as polyfluorenes have high luminescence because they exhibit high absorptions and emissions at the edge of the cited band, as other conductive polymers such as poly(phenylenevinylene). This luminescent character is also related to a delocalization and polarization of the electronic structure. Thus, when the system is excited, the energy is absorbed by some photons to create excitons (electron–hole pairs). This property of the luminescent-conductive polymers is the basis for their application as visual sensory polymers [75–77].

When the chemical structure provides the conductive polymers with higher delocalization and polarization, the luminescence goes down reaching non-fluorescent or weakly fluorescent conductive polymers. This is the case of the first discovered conductive polymer, namely, polyacetylene, in which free carriers are generated from the dissociation of excitons, and they migrate throughout the system. Generally, these free carriers are deactivated by non-radiative processes, but the formation of triplets is also possible. In addition to free carriers, there is another type called local carriers, generated by doping the conductive polymer, which in chemical terms can be cations, anions, radicals and so on. Figure 3.5 shows the chemical structures of the best known conducting polymers, such as polyaniline (PAni) [78], PPy [79], polythiophene [80] and poly(3,4-ethylenedioxythiophene) (PEDOT) [81]. However, there is a wide collection of advanced conductive polymers as sensors: conjugated polymers with synthetic receptors and functional groups (polyalkyl ether and crown ether functionalization conjugated polymers, functionalized with pyridyl-based ligands, enantioselective conjugated polymers, etc.); conjugated polymers with appended protein ligands, nucleotides or DNA; embedded or attached redox-active enzyme-based sensors; induced-fit proteins attached to conjugated polymers and so on [71].

In short, the chemical structure of conducting polymers supposes a great matrix for the preparation of sensory polymers for different targets and with different responses. The most common ones are based on the detection of doping agents (change from weakly conductive to highly conductive), detection of redox species (potentiometric changes) and detection of species implying changes in the luminescence of the system [82–84].

3.3.1.5 Other reversible mechanisms

Other reversible mechanisms give rise to sensory polymers such as ion exchange processes, protonation/deprotonation processes, the formation/breaking of hydrogen bonds or the formation/breaking of weaker bonds such as van der Waals and electrostatic forces.

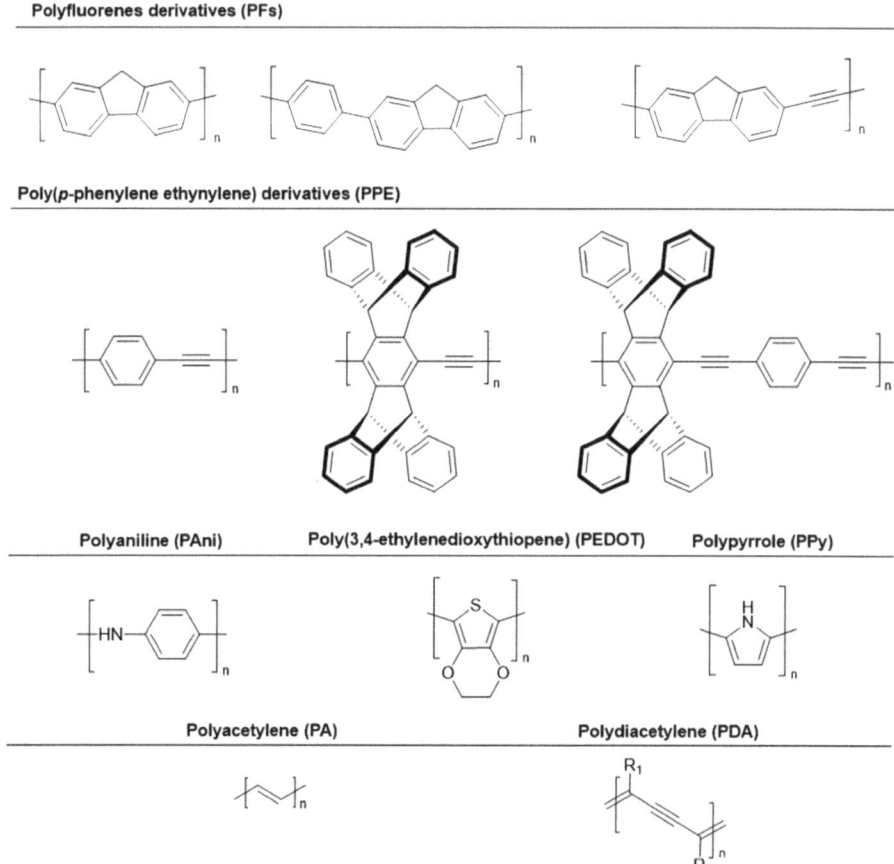

Figure 3.5: Chemical structures of the most common conductive polymers.

These types of mechanisms are especially useful to detect species such as anions of interest, since the exchange of one anion for another in an organic, inorganic or hybrid salt can generate changes in the system large enough to be measured. In fact, there are some examples based on visual changes, as in the case of a chloride anion sensory polymer with direct application in the control and diagnosis of cystic fibrosis, by quantifying these anions in patients' sweat (Figure 3.6) [85].

3.3.2 Irreversible mechanism-based sensory polymers

3.3.2.1 Polydosimeters
Dosimeter-based sensory polymers (polydosimeters) imply an irreversible chemical reaction between the receptor units of the polymer and the target. The formation of

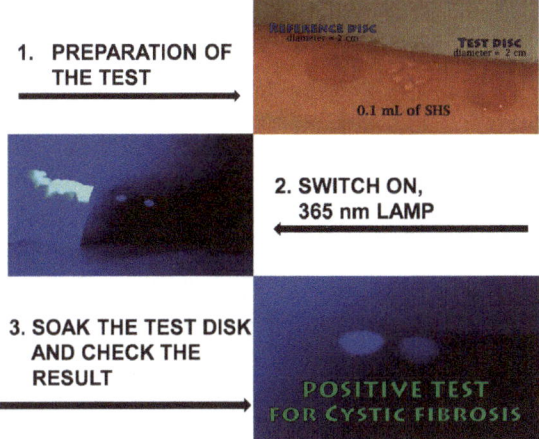

1. PREPARATION OF THE TEST

2. SWITCH ON, 365 nm LAMP

3. SOAK THE TEST DISK AND CHECK THE RESULT

Figure 3.6: Procedure followed for the determination of human chloride in the sweat test. Reproduced with permission of the Royal Society of Chemistry [85].

these new units is traduced in a variation of some property of the system, and the return to the starting point is not possible. These types of sensor polymers are often used to detect chemical species, generating a response that cannot be erased or eliminated. They are also handy for detecting blows (changes in force or pressure) in shipments of fragile or delicate items since they act as a permanent indicator that will attest to a problem regardless of the time that has passed since it.

The potential of polydosimeters has proven to be extraordinary since with them it has been possible to detect species both in the gas phase [51, 86], such as in aqueous media [87–89], as well as by direct contact with a solid or a foodstuff [53].

The appearance of polydosimeters has allowed the creation of new sensory polymers based on chemical reactions discovered and characterized decades ago, giving them a new life by improving aspects such as ease of use or stability. This is the case of the azo-coupling reaction, which occurs between a diazonium salt and a nucleophile as an activated aromatic ring, or an organic anion. Specifically, benzene diazonium salts are prepared through the reaction between an aniline derivative with a nitrosyl cation. These salts are extremely unstable, they decompose quickly after being prepared, and this instability is exacerbated in the solid state. However, when these types of benzene diazonium salts are prepared on a polymer with aniline side groups, they remain stable for at least two weeks, and sensory polymers can be prepared easily and in a simple way for the detection of potentially dangerous species such as phenol derivative-based pesticides [90]. Additionally, the same sensory polymer based on benzene diazonium side groups also finds application as a polyphenol detector in wines [91], or as a detector of antioxidant activity in honeys [92].

A new line of research that is gaining interest in the scientific community is the preparation of polydosimeters through reactions on films with anchorage functional groups (solid or gel-state functionalization). These groups are reactive, and can be chemically modified to obtain sensory motifs in polymers through modification of these anchorage motifs. This new sensory polymer preparation methodology is especially useful when the synthesis of sensory monomers is tedious, or economically unfeasible. Thus, polydosimeters based on ninhydrin sensor motifs can be obtained to detect amino acids, through the oxidation of aminoindanone side groups present in the polymeric structure [49]. In the same way, polydosimeters based on dithizone sensory motifs can be obtained for the detection of Hg(II) through the modification of polymers with aniline side groups present in the polymeric structure, following the synthetic route of Bamberger (Figure 3.7) [93].

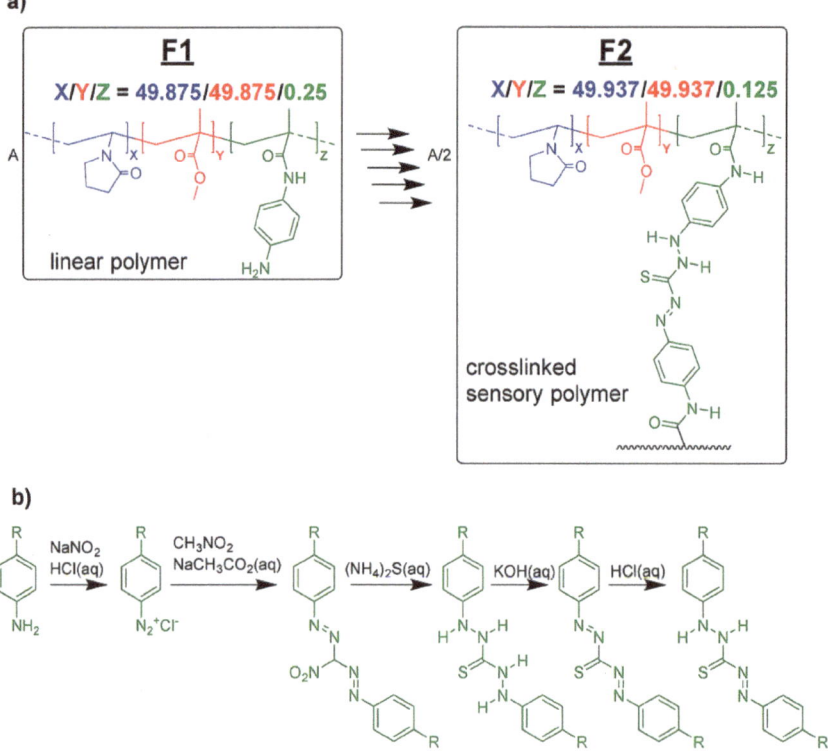

Figure 3.7: (a) Structure and schematic preparation of the sensory films F2 from F1; (b) five-step route to prepare the sensory and cross-linking dithizone-derivative moieties. Reproduced with permission of the Royal Society of Chemistry [53].

Polydosimeters are based on irreversible reactions, but not necessarily on the formation of discrete motifs or discrete molecules. The reactions on which they are based can also give rise to polymers. In fact, some targets such as certain oxidants, are catalysts for polymerizations such as the synthesis of PAni. This fact can be exploited to prepare a sensory polymer containing dispersed aniline to detect and quantify oxidants in the gas phase. The dispersed aniline contained in the sensory polymer starts polymerizing in the presence of targets (oxidants), providing a dual response. In this case, PAni is a coloured conductive polymer, so the formation of this polymer inside the sensory polymer generates changes both in the material's colour and in the conductive properties [94].

3.3.2.2 Polymers containing enzymatically degradable bonds

Fluorogenic and chromogenic peptide-based substrates have been used for many years to analyse the activity of different proteases [95]. In most cases, these peptides can intrinsically be considered polymers since they reach molecular weights around 2,000 by polymerizing tens of amino acids. Generally, the higher the number of amino acids, the higher the specificity of the substrate towards the protease. But these peptide-based substrates are also modified to enable copolymerization with structural monomers to synthesize polymers containing enzymatically degradable bonds [96–101].

In addition to amino acids, the side chain peptide-based substrates contain motifs of fluorophores or chromophores covalently anchored to the peptide. Regarding the first, the fluorophores are combined with quenchers, and give rise to systems based on non-radiative energy transfer processes between a donor and an electron acceptor, specifically, Förster/fluorescence resonance energy transfer processes [102, 103]. In this way, the system only shows fluorescence when an enzyme cleaves the peptide, and separates the fluorophore from the quencher [104].

Other systems are even more straightforward and are based on breaking an amide bond between a fluorophore/chromophore and a peptide to give rise to two fragments, that is, a fluorophore/chromophore with an amine terminal group, and a peptide with a terminal acid group, or vice versa. However, the fundamental principle of the system is the same, as it only exhibits fluorescence when the protease acts on the substrate.

3.3.2.3 Other irreversible mechanisms

There are other irreversible mechanisms through which a sensory polymer can generate a response in the presence of a stimulus. Some of these mechanisms were discovered decades ago, such as enzyme-based sensory polymers, and others are much more recent, such as self-immolative polymers.

The preparation of enzyme-based sensory polymers implies the immobilization of a biological macromolecule on a polymeric substrate, such as enzymes or antibodies. Due to the high relevance of this family of sensory polymers in advanced applications, this book contains an entire section dedicated to the description of enzyme-based sensory polymers, specifically to the immobilization process (**Section 3.6**). This immobilization leads to sensors with a great variety of applications, such as the detection of biomolecules, the preparation of rapid diagnosis kits that is pregnancy tests, SARS-Cov-2 antigen tests and even drug delivery.

In the case of self-immolative polymers, the polymeric chains contain a protected triggering group. In the presence of a target, the triggering group is deprotected and initiates a controlled depolymerization, releasing the building blocks from which it was prepared, and generating a response that allows the quantification of the target [6, 31, 105]. One of the most recurrent examples of this type of mechanism is polyurethanes that contain protected amine groups as triggers [106].

3.4 Classification of sensory polymers by response

Considering the nature of the response generated by the sensory polymers, this section is aimed to classification by the most relevant responses, that is, colorimetric, fluorometric and electrochemical sensory polymers.

3.4.1 Colorimetric sensory polymers

In these sensory polymers the response is a colour change, and it can be recorded with an ultraviolet-visible spectrophotometer, through the digital colour analysis, or even with the naked eye.

Colour is a macroscopic property that is determined by the chemical structure of the system. It depends on the energy difference between HOMO and LUMO. For example, when red colour is observed, green colour is absorbed (780–650 nm), reaching relatively small differences between two cited molecular orbitals. On the other hand, when this difference is higher, dark blue is absorbed (435–480 nm) and yellow is observed.

Colorimetric sensory polymers are the most desired systems for preparing sensors that are easy to use and very intuitive, so that even people without scientific knowledge can use them. In colorimetric sensory polymers, the interaction of the analyte with the target leads to a change in the colour of the material. Moreover, the colour variation is directly related to the concentration of the target species, which allows an easy quantification of the substance in terms of colorimetric parameters. The numerical evaluation of the colour variation was traditionally carried out using a spectrophotometer, and more recently, through the determination of RGB parameters, which can be easily quantified also using a smartphone application [107–109]. After the smartphone boom in 2010, colorimetric sensor polymers have been perfectly complemented with these devices, since colour changes can be easily recorded with the integrated camera, and analysed with very simple software (smartphone applications) that allows it to evolve from a semi-quantitative system (analysis with the naked eye) to a completely quantitative system. Sometimes offering limits of detection similar to spectrophotometric techniques, enough for many applications [107–109].

3.4.2 Fluorometric sensory polymers

Regarding the response, fluorometric sensory polymers are analogous to colorimetric ones. Fluorescence is related to radiative transitions from an excited electronic state to the fundamental one emitting radiation. Depending on the type of transition, two phenomena could occur, fluorescence or phosphorescence, which is explained with the Jablonski diagram [110].

The response of this type of sensory polymer is recorded with fluorometers (Figure 3.8), but in recent years, digital colour analysis enhanced by smartphones has also been used [85]. However, unlike colour changes, fluorescence changes are not visible to the naked eye, and the sample must be illuminated with UV light (wavelengths ranging from 250 to 400 nm).

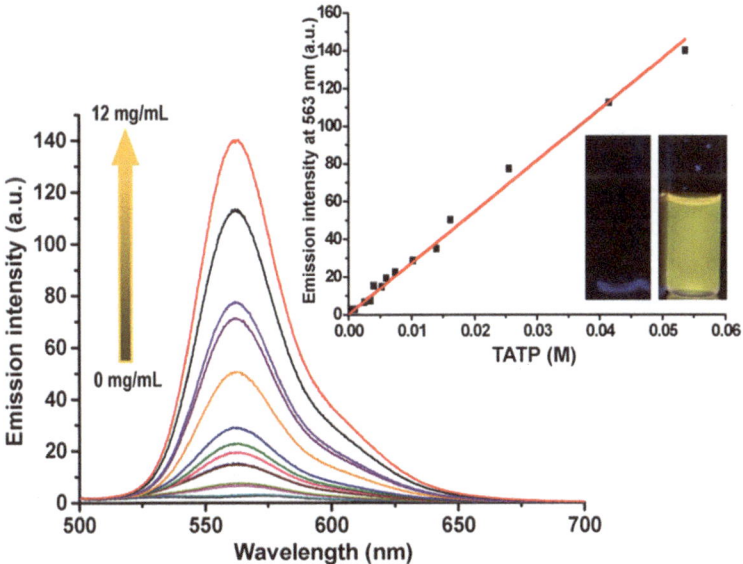

Figure 3.8: Variation of fluorescence curves of the sensory polymer (left) and titration plot (right) against the solution concentrations of TATP. It is also clearly visible with the colour change of the polymer dissolution when TATP is detected. Reproduced with permission of Wiley [111].

However, it is possible to take digital photographs once the sample is illuminated, which ultimately means a translation of fluorescence into parameters of a digital colour space, such as RGB, HSV and CIE-LAB space [112].

3.4.3 Electrochemical sensory polymers

Electrochemical polymeric sensors are also known as chemoresistive sensors, and they can be schematically represented as in Figure 3.9. This kind of sensory polymers are commonly based on conductive polymers, sometimes combined with MIPs [113–115].

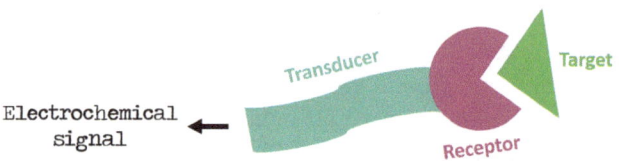

Figure 3.9: Schematic presentation of an electrochemical sensor.

They transform the chemical information into an electrical response, and a broad range of electrochemical techniques can be used to analyse this response. The most common ones are voltammetry (measurement of the system's current when modifying potential), amperometry (a uniform potential is applied and the change in current is monitored as a function of time), potentiometry (measurement of the system's potential under no current flow), impedometry (application of an alternating potential and follow up of the perturbation in the steady state) and conductometry (measurement of the ability of the system to conduct an electrical current).

There are two main applications in which chemoresistive sensory polymers are employed: gas sensing and heavy metal ions detection. Gas sensing mechanisms involve the charge conduction mechanisms achieved through intra- and interchain transports [116]. Different types of charges are present in the conducting polymer, namely polarons, bipolarons or solitons, which introduce energy states between the LUMO and the HOMO. Then, electrical resistance variation is directly related to the presence of these charges and their mobility. Thus, the mechanisms of the creation of these charges define the electrical response of chemiresistive sensors.

In the case of the interaction with gas substances, the resistance variation depends on the nature of both the sensing polymer and the gas molecule. For a p-type semiconducting polymer, a determined quantity of the oxygen of the air is continuously adsorbed on the surface, removing the electrons from the conduction band. This process led to creating single or double oxygen ions, which the surface cannot adsorb. Then, when a reducing gas such as ammonia (NH_3) reacts with these unadsorbed species, hole concentration is reduced, and resistance increases. On the other hand, in the presence of an oxidizing gas such as nitrogen dioxide (NO_2), electrons are consumed from the valence band, increasing hole concentration and decreasing resistance. The sensing mechanism is inverse in the case of n-type conducting polymers. The p-type or n-type behaviour of conducting polymers depends on the doping procedures. Then, doping with oxidizing agents (*p*-type doping) introduces charge carriers into the electronic structure, forming a *p*-type conducting polymer and vice versa [117]. Figure 3.10 shows the sensing mechanism of a *p*-type material in air and reducing gas environments, whereas Table 3.1 indicates the sensing response of *p*-type and *n*-type sensors (adapted from Ref. [118]).

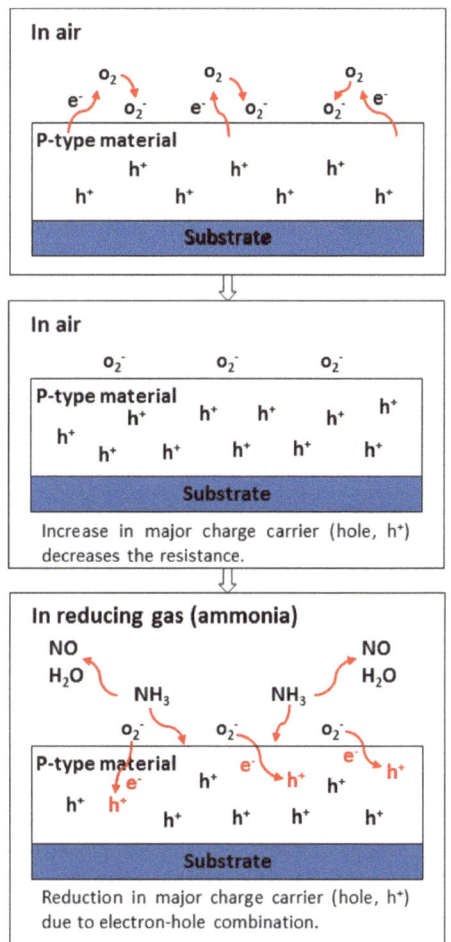

Figure 3.10: Sensing mechanism of p-type conducting polymer in air and reducing gas environments. Reproduced with permission of: IOPscience [118].

As seen in Table 3.1, many different gases can be detected using chemiresistive sensors, although ammonia is the most analysed gas [119]. PPy-based sensors present a very similar response (in terms of *R*-value) for a wide range of concentrations (20 to 10,000 ppm), although this parameter can reach higher values (up to 2.5), when higher concentrations are present [120]. Other gases can also be detected, such as

H_2, NO_2 or methanol [121]. In these cases, the better response ($R \approx 24$) is obtained when PPy nanosheets are employed [122]. PAni has been employed to detect ammonia gas with values of R ranging from 1 to 100 in concentrations between 20 and 1,000 ppm [123, 124], and it has also been used in the detection process of other gases (CO, NO_2, Cl_2, etc.), but the responses showed almost equal R values in a wide range of concentrations, except in a very few cases in which thin films were fabricated [125]. Other conducting polymers that can act as chemiresistive sensors are polythophene and PEDOT. Their response was highly dependent again on the material's geometry, obtaining promising results in the case of using poly(3-butylthio) thiophene) films in the detection of NO_2 ($R \approx 16$) [126] and PEDOT nanowires to detect nitric oxide ($R \approx 7$) [127].

Table 3.1: Sensing response of p-type and n-type chemoresistive sensors.

Sensing response behaviour	p-Type sensor	n-Type sensor	Analytes
Reducing analytes	Resistance increases	Resistance decreases	CO_x, NH_3, CH_4, H_2, H_2S, methanol
Oxidizing analytes	Resistance decreases	Resistance increases	NO_x, CO_2, SO_2, O_2, O_3
Dominant charge carrier	Holes (h^+)	Electrons (e^-)	N/A

The use of electrochemical polymer-based sensors in the detection of heavy metal ions has emerged as an alternative to classical methods (atomic absorption spectroscopy, X-ray fluorescence spectrometry and so on [128]) due to the capability of performing fast analysis combined with high sensitivity and simplicity [129]. Among all the heavy metal ions, mercury has been the most analysed in the last years, using mainly a modified electrode in which conducting polymers such as PAni or PEDOT were deposited onto its surface using different electric measurements (amperometry, potentiometry or cyclic voltammetry) to detect the presence of the analyte [130]. Recently, Wang et al. have presented a review about the use of chemiresistive polymers in heavy metal ion detection [131]. The importance of conducting polymers such PPy, PEDOT or PAni in this field is increasing greatly, although in most cases, these SPs are modified or synthesized as a part of more complex polymeric systems. PPy has been functionalized with iminodiacetic acid (containing carboxylic groups) in a modified electrode to detect different heavy metal ions (Pb(III), Hg(II), Cd(II) or Co(II)) [132], traces of Hg(II) were detected using nanocomposites based on graphene oxide/ PEDOT [133], and Deshmukh reported the detection of Cu(II), Pb(II) and Hg(II) using EDTA chelating ligands to modify PAni and single wall carbon nanotubes [134]. On the other hand, it is important to remark the importance of MIPs in this research field, because of their advantages as strong selectivity, simple synthesis procedures and the possibility of building appropriate recognition sites. In this sense, different literature works can be found in which Pb(II) or Cu(II) are detected and quantified at trace level [135, 136].

3.5 List of the most relevant sensory polymers sorted by application

The number of applications related to sensory polymers is extremely extensive, and it goes beyond the scope of this book to review them in detail, so a list of the most relevant examples classified according to their application are given in Table 3.2, together with the reference for further information.

Table 3.2: List of the most relevant sensory polymers sorted by application.

Application		Stimulus	Mechanism	Response	Reference
Biomedicine	Amino acid detection	Chemical – liquid phase	Reversible – MIP	Colorimetric	[137]
	Escherichia coli detection	Chemical – liquid phase	Reversible – MIP	Electrochemical	[138]
	Chronic wounds monitoring	Chemical – liquid phase	Irreversible – polydosimeter	Colorimetric	[49]
	Cystic fibrosis monitoring	Chemical – liquid phase	Reversible – ion exchange	Fluorimetric	[85]
	Iron detection in blood	Chemical – liquid phase	Reversible – organometallic complex formation	Colorimetric	[87]
	Detection and inhibition of bacteria	Chemical – liquid phase	Irreversible – polydosimeter	Fluorimetric	[139]

Table 3.2 (continued)

Application		Stimulus	Mechanism	Response	Reference
Food control	Determination of chlorpyrifos in apple juice	Chemical – Liquid phase	Reversible – MIP	Colorimetric/ Raman	[140]
	Determination of melamine in milk	Chemical – liquid phase	Reversible – MIP	Electrochemical	[141]
	Phycocyanin detection	Chemical – liquid phase	Reversible – fluorescence quenching	Fluorimetric	[142]
	Determination of antioxidant activity in honey	Chemical – liquid phase	Irreversible – polydosimeter	Colorimetric	[92]
	Determination of total polyphenol index in wines	Chemical – liquid phase	Irreversible – polydosimeter	Colorimetric	[91]
	Determination of mercury in fish	Chemical – solid phase	Reversible – organometallic complex formation	Colorimetric	[53]
Soil, water and air pollution	Cyanide detection	Chemical – liquid phase	Irreversible – polydosimeter	Colorimetric	[143]
	Bisphenol A detection	Chemical – liquid phase	Reversible – MIP	Colorimetric	[144]
	CO_2 detection	Chemical – gas phase	Reversible – conducting polymer	Electrochemical	[145]
	NO_2 detection	Chemical – gas phase	Reversible – conducting polymer	Electrochemical	[146]
	CO detection	Chemical – Gas phase	Reversible – Conducting polymer	Electrochemical	[147]
	Pesticide control	Chemical – liquid phase	Irreversible – polydosimeter	Colorimetric	[90]

Table 3.2 (continued)

Application		Stimulus	Mechanism	Response	Reference
Civil protection	Nitroexplosive-Picric Acid detection	Liquid and solid phase	Reversible – conducting polymer	Fluorimetric	[148]
	TNT detection	Liquid and gas phase	Irreversible – polydosimeter	Colorimetric	[149, 150]
	TATP detection	Liquid and gas phase	Reversible – quenching process	Fluorimetric	[111]
	Oxidant atmospheres detection	Chemical – gas phase	Irreversible – polydosimeter	Colorimetric/ electrochemical	[94]
	Extreme acidity	Chemical – gas phase	Reversible – protonation/ deprotonation	Colorimetric	[151]
	Acidic atmospheres	Chemical – gas phase	Reversible – protonation/ deprotonation	Colorimetric	[152]

3.6 Immobilization of biomolecules

3.6.1 Introduction

The first phenomenon in the detection of a target molecule by a chemical sensor is the selective interaction between the analyte and the sensor; the recognition and affinity of the sensor towards the analyte are brought to light. In general, if the recognition is based on biochemical or biological nature, the term biosensor is used instead of chemosensor. One of the most common recognition methods taking part in these processes is the detection through a reversible weak interaction, non-covalent, between the biosensor and the analyte. The most representative example is the interaction of an antigen with its antibody, which is highly selective. Antibodies are glycoproteins produced by the immune system due to the presence of an antigen (caused by virus, pathogen bacteria, etc.) [153, 154, 163–165, 155–162].

The immobilization of biomolecules such as antibodies, antibody fragments or enzymes, is of great interest for the detection of other biomolecules such as antigens, or peptide substrates. In fact, the antigen/antibody interaction, or the recognition of a peptide substrate by an enzyme, are just two examples of detection systems present in the biological environment, which sciences such as chemistry, biology, biotechnology and materials science attempt to mimic when immobilizing a biomolecule on a solid support.

The most obvious application of biomolecule immobilization techniques is the preparation of systems able to detect species of interest, such as another biomolecule related to a disease, that is, an antigen. In this way, by immobilizing a protein (antibody, antibody fragment or enzyme) on a solid support, a system able to detect another protein (antigen, peptide, etc.) is generated.

Antibodies are immunoglobulins (Ig), a type of proteins of about 150 kDa and sizes about 14 x 10 x 4 nm. They show a three-dimensional (3D) structure defined by a precise amino acids sequence. Figure 3.11 shows the basic structural unit of an antibody, constituted by four protein chains, two light chains (L) of about 25 kDa and two heavy chains (H) of about 50–70 kDa each one depending on the type of immunoglobulin, bind together through non-covalent bonds and disulphide bonds, forming the quaternary structure of the protein. LC are formed by two domains and HC by four. Antibodies have two types of domains, constants (C) and variables (V). Constant domains show an amino acid sequence similar for all the antibodies of the same type, while the sequences of the *N*-terminal domain of heavy and light chains (named HV and LV) are variable among the different types of antibodies and they constitute the variable fragment (VF), where the antigen interacts with the antibody. The former domains of the chain, H and L constitute the Fab fragment (antigen-binding fragment), while the latter constant domains (HC) of both H chains constitute the crystallizable fragment (Fc). This region of the antibody modulates the immune activity through the binding to specific receptors, what ensures that each antibody generates an immune answer to a specific antigen. Antibodies show carbohydrates chains in the Fc region bound to the amide nitrogen of an asparagine residue, being, as a result, glycoproteins.

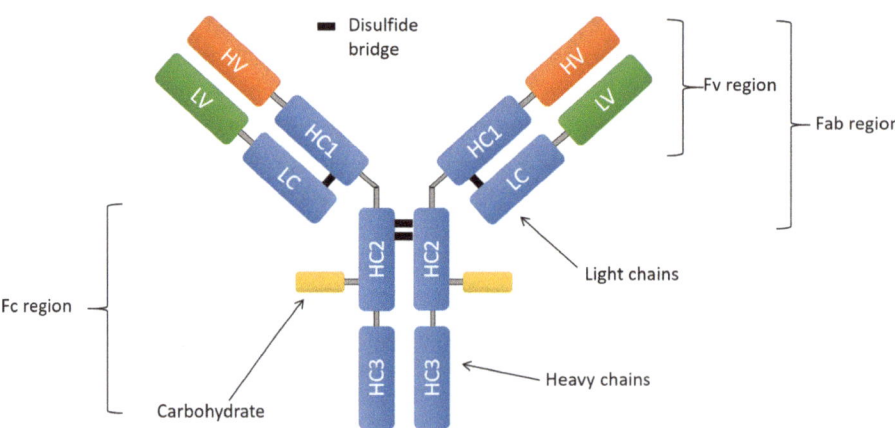

Figure 3.11: Schematic representation of an antibody.

Undoubtedly, the best-known technique based on this principle is the enzyme-linked immunosorbent assay (ELISA) technique, which emerged as an alternative to the radioimmunoassay, in which harmful radioactive isotopes were used. Also, rapid tests to detect SARS-CoV-2 antigen quickly and cheaply were developed following the same principles. Furthermore, pregnancy tests, tests to detect allergens in food or even some tests to detect viruses are based on the ELISA technique, highlighting the importance of the immobilization of antibodies or enzymes on solid supports.

An adequate immobilization of a biomolecule to solid support ensures a correct fabrication and performance of a biosensor. A preferable orientation avoids denaturalization and keeps the active site exposed to maintain the affinity towards another molecule. The active site needs to be in a specific conformation and accessible to the target species, while random orientation can denaturalize and block the active site. Traditional physical immobilization methods involve the adsorption of the proteins to the substrates through electrostatic forces of hydrophobic interactions. These strategies lead to unstable and uncontrollable immobilization, which reduces sensitivity. Covalent attachment through functional groups present in the biomolecule with modified substrates is irreversible and efficient, but it is also non-oriented.

Oriented protein immobilization is important for bioanalysis, and various strategies are being carried out. In this sense, polymers play an important role in the immobilization of biomolecules, which is another reason they are called "smart". Regarding the immobilization of biomolecules, SPs can be divided in four types, schematically represented in Figure 3.12:

- Type 1 SP. The polymer or copolymer acts as support and attachment point; that is, it makes the final product manageable and allows, from a chemical point of view, the attachment of a protein, either by covalent bonds, or by affinity, by entrapment and so on.
- Type 2 SP. They are formed by a Type 1 SP and a polymer placed between the biomolecule and the Type 1 SP. This polymer makes the chemical or physical attachment of the support and the biomolecule possible. These polymers are also used as chain spacers, functionalization agents or even orientation agents. For example, if the biomolecule to be immobilized is an antibody (usually represented as "Y" due to its shape), it can orient the attachment of the antibody in the right way.
- Type 2 *pseudo*-SP. They have the same composition as Type 2, but instead of using a polymeric Type 1 SP, a non-polymeric support is used, such as an electrode, a glass support or a metal support.
- Type 3 SP. They are formed by a Type 2 SP and a protein, aimed to an almost perfect orientation of the immobilized biomolecules (in this case, antibodies).

Some types of SP may seem very simple, but they need to place macro-biomolecules as antibodies, with sizes around 150 KDa, and approximate dimensions of 14 x 10 x 4 nm. SPs make this possible for several reasons: they can provide many and different functional groups prone to form covalent bonds with antibodies; different polymeric structures can be selected according to the needs of the final applications, making it possible to control the surface properties to immobilize antibodies specifically

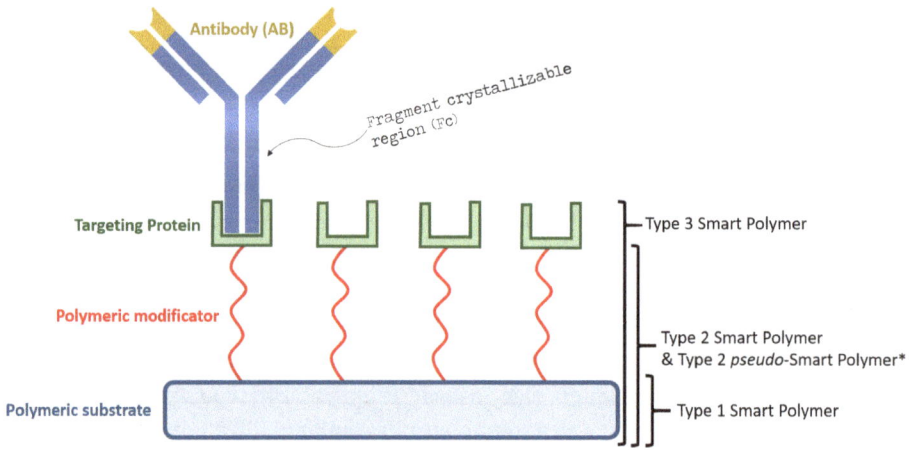

Figure 3.12: Schematic representation of different types of smart polymers. (*) Type 2 *pseudo*-smart polymers are composed by a non-polymeric substrate instead of a Type 1 smart polymer.

and to diminish the immobilization of other proteins; they can locate long chains and proteins on the surface to act as spacers between the substrate and the antibody and thus favouring the immobilization of antibodies and their interaction with antigens.

Furthermore, for the right functioning of the detection system, the antibodies must be correctly positioned and attached. In other words, the orientation of the antibody must be appropriate, and the interaction with the SP must be robust, which means that they should not break apart under normal use conditions. Thus, there are two important characteristics of SPs that must always be considered: orientation and linkage.

3.6.1.1 Orientation

SPs favour the ideal orientation of antibodies. Thus, if an antibody is immobilized on a "non-smart" surface, the orientation would be random. There will be lying down antibodies (or "lying-on"), leaning antibodies (or "side-on"), upside-down antibodies ("or head-on") and suitable oriented antibodies (or "end-on"). Figure 3.13 shows antibodies randomly immobilized on a "non-smart" surface.

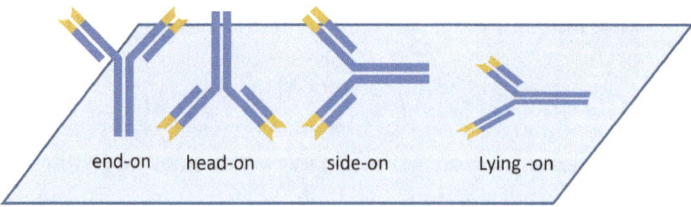

Figure 3.13: Schematic representation of randomly immobilized antibodies on a "non-smart" surface.

There are different strategies to favour the appropriate orientation of antibodies. One of them is through the disulphide groups of the antibody. One of the most straightforward is the use of Type 1 SP containing phenylboronic acids in their formulation since they favour the "end-on" orientation due to the reaction of the diol groups of the glycosylate in the crystallizable fragment of the antibody. To characterize the succeed of this orientation procedure, Time-of-flight secondary ion mass spectrometry technique is carried out, which is used to analyse the composition of solid surfaces and thin films by spraying the surface of the sample with a beam of focused primary ions, and collecting/analysing the ejected secondary ions.

In an ideal situation, the perfect SP would orient 100% of the antibodies in the right and accessible position (end-on), but so far, this does not occur. However, as described later regarding the linkage of polymers and antibodies, very high percentages are achieved with Type 1 SP and Type 2 SP, and especially with Type 3 SP.

3.6.1.2 Linkage

The immobilization process can be carried out physically (adsorption, electrostatic interactions, affinity) or chemically (covalent bond). Each of these ways has its advantages and disadvantages. For example, physical methods do not modify the structure of the immobilized biomolecule; they simply encapsulate it or retain it through electrostatic interactions. This is a great advantage since a modification in the biomolecule structure can lead to its denaturation or deactivation. However, the interactions between the SP and the protein are weaker than in the case of chemical attachment, which sometimes results in the release of the protein from the substrate. Furthermore, in physical linking carried out by encapsulation, the interaction of the attached antibodies with the targets (antigens) can be more impeded and, depending on the SP's final application, which can be a drawback.

The use of 3D substrates, MIPs, electrostatic interactions (layer-by-layer materials, LbL) and taking advantage of chemical polarity (Langmuir–Blodgett (LB) materials) are the physical linking methods most used currently.

On the other hand, chemical methods generate covalent bonds of different types, but all of them are very resistant and reliable. However, depending on the area where the antibody binds the substrate and how specific these bonds are, the antibodies could lose their activity. The antigen/antibody interaction is so specific and so delicate that any change in the chemical structure of both can lead to the inactivation of the system. That is precisely why SPs have to be designed appropriately and ad hoc for each application.

In the case of a covalent bond between the support and the antibody, a *sine qua non* condition for Type 1 SP is that they have functional groups to enable chemical reactions and thus succeed in the immobilization of antibodies. Polystyrene (PS) is not a SP, and in fact, it is a fairly inert material. As depicted in Figure 3.14, it is a linear polyethylene chain with benzene lateral groups from the chemical point

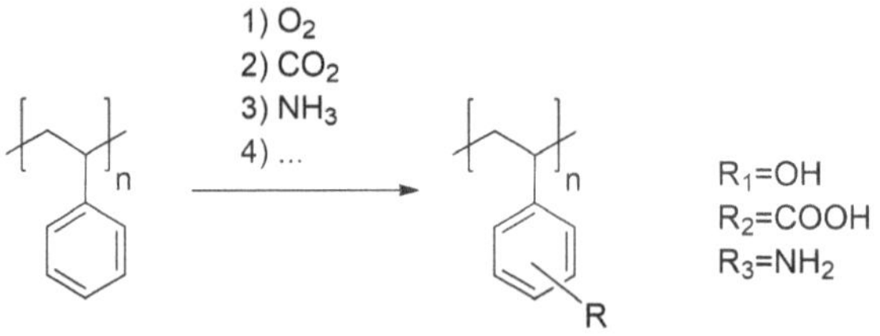

Figure 3.14: Generation of different functional groups on the surface of an inert polymer as polystyrene by using plasma.

of view. It does not have any functional groups capable of generating any type of chemical reaction. However, with the appropriate treatment and design of materials, PS can become a Type 1 SP. These treatments generate modifications in the material, which allow the inclusion of functional groups susceptible to a chemical reaction. These modifications are further discussed in this chapter.

3.6.2 Smart polymers for the immobilization of biomolecules

3.6.2.1 Type 1 smart polymers

Type 1 polymers are made by just one polymer substrate, which can be modified (or not) with different techniques: surface modification with plasma for polymers such as PS, poly(vinylidene fluoride), PMMA, polydimethylsiloxane (PDMS); taking advantage of click reactions; using 3D substrates; through MIPs; through controlled polymerization techniques such as atom transfer radical polymerization (ATRP) or reversible addition-fragmentation chain-transfer polymerization. For Type 1 SP, the attached biomolecule is directly linked. The polymer acts as support material, and the immobilization can be carried out by both physical and chemical methods. The different Type 1 SP can be classified according to their manufacture or preparation techniques as follows.

a) Preparation of Type 1 smart polymers by using plasma and functionalized polymers

Plasma is considered the fourth state of matter and it is generated using gases (such as CO_2, O_2, N_2, NH_3 and H_2) and an energy source such as microwaves, radio waves or high-energy electrons. This technique generates functional groups on the surface of an originally inert polymer, such as PS. This allows both the direct anchoring of antibodies to prepare a Type 1 SP and its further modification with another polymer for the preparation of Type 2 SP [156].

Thus, depending on the gas and the experimental conditions, carboxylic, amino, methyl or hydroxyl groups can be generated on the surface of the inert polymer.

Additionally, this process also increases the affinity of the material for aqueous media, making it more hydrophilic and avoiding the denaturation of the immobilized biomolecules that usually occurs when wholly hydrophobic supports are used. Taking PS as an example, Figure 3.14 represents the generation of functional groups with different gases [162].

For PS, the plasma modifications are most likely to occur in the aromatic ring, since it is the most susceptible to modifications moiety. However, the generation of plasma with a gas can also involve modifications in the main chain of the polymer, which explains why polymers such as PTFE, or PMMA can be used for the preparation of Type 1 SP.

Other polymers can intrinsically contain these types of functional groups in their structure, so plasma modification is not necessary, and they are used directly as Type 1 SP. Examples of these polymers are polylysine or polyethylenimine containing $-NH_2$ functional groups. Also, carboxymethylcellulose has $-COOH$ groups, and polymers widely known as polysaccharides contain $-OH$ groups. The methodology for the formation of new bonds described below can be used indistinctly for polymers containing generated or intrinsically present functional groups.

Once Type 1 SP is prepared, the next step is to bind it to a biomolecule (as an antibody) and thus immobilizing it. In the case of Type 1 SP with $-NH_2$ groups, the reaction is carried out with free $-COOH$ groups, due to the presence of amino acids such as aspartic acid or glutamic acid present on antibodies along their structure. This type of reaction is commonly referred to as "carbodiimide chemistry" that uses N-ethyl-N-(3-dimethylamino propyl) carbodiimide (EDC) combined with succinimidyl esters such as N-hydroxysuccinimide (NHS). This chemistry is known as EDC/ NHS coupling and results in the formation of solid amide bonds between the antibody and the Type 1 SP. This methodology is also used with Type 1 SP modified with $-COOH$ groups, with the reaction taking place through the free $-NH_2$ groups present in the proteins and provided by amino acids such as lysine. Figure 3.15 schematically shows these procedures on a generic inert polymer, such as PS or PTFE. However, this figure shows an ideal orientation of the antibody on the Type 1 SP surface, which does not represent the real situation, in which random orientation occurs in these conditions.

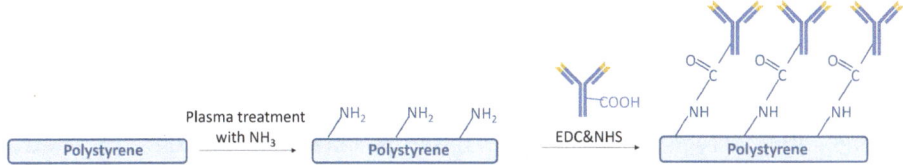

Figure 3.15: Schematic representation of the attaching process of an antibody on to the surface of a Type 1 SP by EDC/NHS chemistry.

An alternative to the reaction between acids and amines is the reaction between thiol groups and maleimide groups. To obtain a Type 1 SP with thiol groups, ammonia plasma is generally used to generate –NH$_2$ groups in the inert polymer surface, and then these groups are transformed into thiol groups with 2-iminothiolane (Traut's reagent). The Type 1 SP could then be linked to a biomolecule modified with maleimide groups. On the other hand, a Type 1 SP could be prepared with maleimide groups (or maleimide derivatives) based on the same chemistry. The procedure is also based on the generation of –NH$_2$ groups in the inert polymer with ammonia and plasma, but this time, these groups are reacted with reagents such as sulphosuccinimidyl-4-(N-maleidomethyl)-cyclohexane-1-carboxylate, or sulfo-succinimidyl -4-(p-maleimidophenyl) butyrate. Figure 3.16 schematically shows these procedures on a generic inert polymer, such as PMMA. Once the Type 1 SP with maleimide groups is generated, a previously modified antibody can be anchored. In this case, the existing disulfide bridges in the Fc region of the antibodies need to be reduced, and this process can be carried out with reagents such as tris-(2-carboxyethyl) phosphine, 2-mercaptoethylamine or dithiothreitol.

The combinations of the selected inert polymer, the gas, and the type of antibody attachment are numerous, and the choice would depend on the final application of the SP and the working conditions.

Table 3.3 summarizes some examples of polymer modification and their features.

b) Preparation of Type 1 smart polymers by using click reactions

Click reactions are a very efficient alternative for the preparation of Type 1 SP through the reaction between an azide and a triple bond. In this sense, Type 1 SP with alkynyl (triple bond) groups are obtained from other Type 1 SP that can be purchased commercially, such as PEG modified with NHS esters. These functionalized PEGs are reacted with propargylamine in CH$_2$Cl$_2$ at room temperature, and the alkynyl-PEGs are obtained (Figure 3.17) [166]. In this case, unlike –NH$_2$ groups or –COOH groups, proteins do not contain azide side groups, so they need to be generated. The most common procedure to render azide groups is based on substituting methionine (natural amino acid) with azidohomoalanine (AHA).

Once the Type 1 SP with a triple bond has been generated, and once the protein has been modified with AHA, immobilization can be carried out through the 1,3-dipolar cycloaddition reaction of Huisgen, in the presence of a catalyst Cu(I), which is known as "click" reaction.

c) Preparation of Type 1 smart polymers based on molecularly imprinted polymers

MIPs are synthetic receptors for a target molecule. They are prepared using the molecular imprinting technique, which leaves cavities in the polymer matrix with affinity for a chosen template molecule, such as an antibody. One of the most widely used polymers is poly(hydroxyethyl methacrylate) (PHEMA) in the form of a gel or

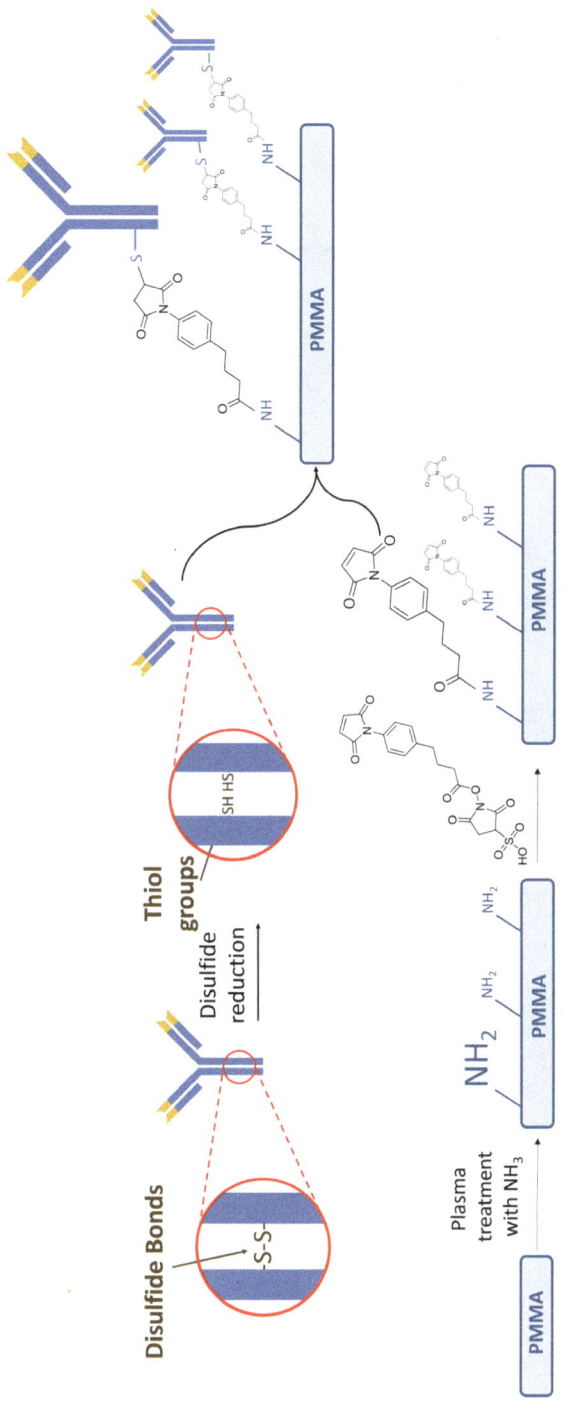

Figure 3.16: Schematic representation of the attaching process of an antibody on to the surface of a Type 1 SP by thiol/maleimide chemistry.

Table 3.3: Plasma treatment of different polymers with different gases and the obtained result [156].

Polymer	Plasma treatment	Result
Polytetrafluoroethylene (PTFE)	N_2 or O_2	Enhanced hydrophilicity
Poly(methylhydrogensiloxane-*co*-dimethylsiloxane) (PMHS-co-DMS)	$(SF_6 + O_2)$ or $(CF_4 + O_2)$	Brittle columnar structure
Low-density polyethylene (LDPE)	Poly(2-ethyl-2-oxazoline), low temperature plasma	Desired wettability and adhesion properties
Polystyrene (PS)	Ar, O_2	Improvement of adhesion properties.
	O_2	Hydrophilic surface
	CF_4	Hydrophobic surface
Polycaprolactonediol (PCL)	$Ar + H_2$	Hydrophobic surface
	$Ar + N_2/Ar/Ar + O_2$	Hydrophilic surface
Poly(lactic acid) (PLA)	Dry air/Ar/He/NH_3	Hydrophilic surface

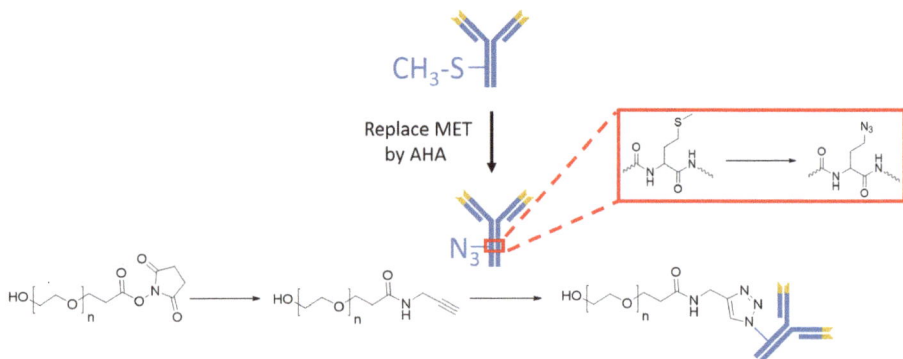

Figure 3.17: Schematic representation of the attaching process of an antibody into a modified PEG (Type 1 SP) through click chemistry.

cryogel. With this type of polymers, the orientation of the antibodies is improved, and no chemical modification takes place in the immobilized biomolecule since it is physically attached. Due to the hydrophilic nature of polymers such as PHEMA, the probability of denaturation of the antibodies is also reduced. Figure 3.18 shows the preparation of a Type 1 SP with PHEMA [167].

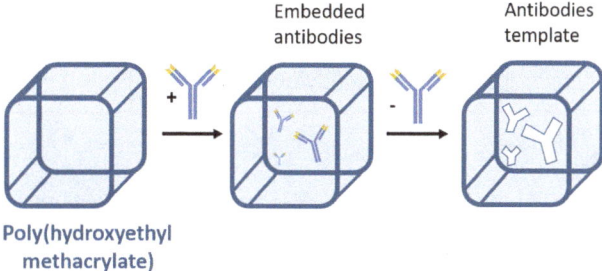

Poly(hydroxyethyl
methacrylate)

Figure 3.18: Schematic representation of the preparation process of a SP based on MIPs.

d) Preparation of Type 1 smart polymers based on three-dimensional substrates

Unlike MIPs, 3D substrates are porous polymeric supports that form a 3D network with cavities that can be used for the physical immobilization of antibodies. The most notable examples are agarose or dextran gels [167], for example, which have improved the attaching process of antibodies due to their high surface area. These two examples are probably the most widely used porous polymers within this group, but 3D immunochromatographic nitrocellulose membranes have also been used [168]. Figure 3.19 schematically shows a 3D substrate for the immobilization of antibodies.

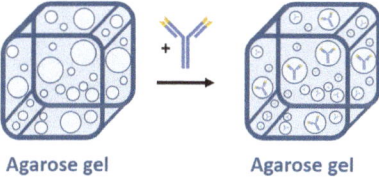

Agarose gel   Agarose gel

Figure 3.19: Schematic representation of a 3D-substrate.

e) Preparation of Type 1 smart polymers based on atom transfer radical polymerization

ATRP is a controlled radical polymerization method, which was developed with the aim of having control over molecular weight and molecular weight distribution of synthesized polymers. It is sometimes called "living polymerization" because grafts of other polymers can be grown in a controlled way. The ATRP method can be used to synthesize Type 1 and Type 2 SP. When Type 1 SP is synthesized using ATRP, the ATRP initiators ("living" polymers) allow further polymerization with another polymer to form a Type 2 SP.

For this Type 1 SP preparation, it is common to use chloromethylated polyethersulphone membranes. They can then react with polymers such as poly(2-(methacryloyloxy)ethyl dimethyl-(3-sulfopropyl)ammonium hydroxide), to prepare a Type 2 SP for the immobilization of biomolecules. Polysiloxanes are also often prepared using

ATRP (Type 1 SP). The "living polymer" can initiate a second polymerization on its structure, by adding monomers (or mixtures of monomers) as *N*-isopropylacrylamide and 2-methoxyethyl methacrylate to the media, to finally obtain a Type 2 SP.

3.6.2.2 Type 2 smart polymers

Type 2 SP combine a Type 1 SP with another polymer that improves the system, in terms of affinity, chain separation, or orientation of the antibody.

a) Preparation of Type 2 smart polymers based on plasma modifications and 3-aminopropyltriethoxysilane

The use of plasma and a gas is still a good strategy when preparing these SPs. In the first step, an inert polymer (such as PDMS) can be modified with plasma and oxygen, generating hydroxyl groups on the material's surface. In a second step, the material is reacted with 3-aminopropyltriethoxysilane (APTES) as described in Figure 3.20. In this reaction, one of the hydroxyl groups of the APTES is condensed with the hydroxyl groups on the surface of the material [157].

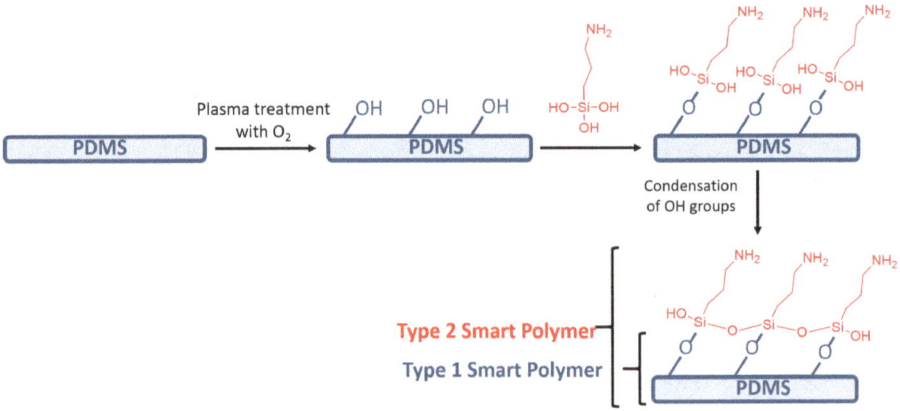

Figure 3.20: Schematic representation of the preparation of a Type 2 smart polymer based on plasma modified PDMS and 3-aminopropyltriethoxysilane (APTES).

The rest of the hydroxyl groups of APTES remain free and can self-condense, producing a polymerization on the material's surface and leaving free $-NH_2$ groups on the surface. At this point, the Type 2 SP is susceptible to immobilize an antibody following different chemical and physical strategies:

- *Adsorption:* The generation of $-NH_2$ groups turns the material's surface hydrophilic, which enables the immobilization of antibodies. However, this strategy produces randomly oriented immobilized antibodies, which is undoubtedly a major disadvantage.

- *Amide formation reactions:* As described previously for Type 1 SP, the amines generated on the surface can react with the acid side groups of certain amino acids present in biomolecules such as antibodies, through EDC/NHS chemistry. Orientation is enhanced compared to adsorption immobilization, but it still can be further improved.
- *Imine formation reactions:* $-NH_2$ groups on the material's surface can be transformed into imine groups, using reagents such as glutaraldehyde. This bifunctional aldehyde reacts with the amines of the material through one of its aldehyde groups. In this way, the other aldehyde group is free, and the material's surface is aldehyde-modified. The last step consists of the reaction between this aldehyde group and the free amine groups present in the antibody, resulting in an imine bond.
- *Oxidation in glycols:* Another preparation strategy consists of the oxidation of the hydroxymethyl groups present in antibodies to aldehydes. Polysaccharides of specific antibody regions provide these hydroxymethyl groups. In addition, oxidizing agents such as periodate are often used to enable the reaction of these aldehydes with the amine groups present on the surface of the material.
- *Thioureas formation reactions:* 1,4-phenylenediisothiocyanate is a molecule that contains an aromatic ring with two isothiocyanate groups. These groups react with amines to form thiourea bonds. This way, using an excess of 1,4-phenylenediisothiocyanate, only one of the groups reacts with the amine groups of the surface, functionalizing the material with isothiocyanate groups. When immobilizing an antibody, these groups can react with the amine groups present in the antibodies (provided by amino acids such as lysine) to form the final thiourea bonds, and immobilizing the antibodies.

b) Preparation of Type 2 smart polymers based on plasma modifications and boronic acids

Boronic acids are compounds that can form boronic esters with the diols present in specific sites of antibodies. These diols are located in the polysaccharides of the crystallizable region of antibodies, which is commonly called Fc. This reaction can be used to anchor antibodies to Type 2 SP. However, firstly the Type 2 SP must contain boronic acids in its structure.

A proven strategy is based on preparing a Type 1 SP using plasma to generate $-COOH$ groups on the surface. These groups will be used to attach a polymer with boronic acid groups (Figure 3.21), which also contains an $-NH_2$ group and is susceptible to forming an amide bond through EDC-based chemistry [169].

The polymer shown in Figure 3.21 holds a third functional group based on 3-(trifluoromethyl)-3H-diazirine. This group is a cross-linking agent that is activated with ultraviolet light, and that generates a cross-linked polymer to improve the workability of the material and the anchoring of antibodies.

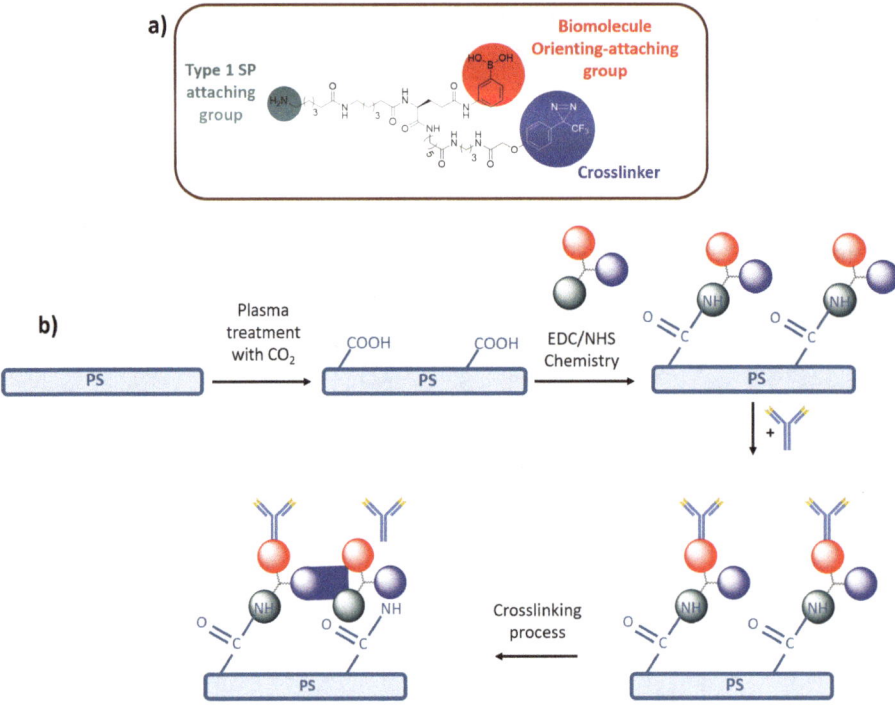

Figure 3.21: (a) Polymer containing −NH₂ groups (for the formation of amide bonds, and the consequent attaching of the polymer to the Type 1 SP), boronic acid groups (for the attaching of the antibodies throughout the diols of the antibodies Fc region), and 3-(trifluoromethyl)-3H-diazirine (for the cross-linking of the system); (b) schematic procedure for the preparation of Type 2 SP with boronic acids.

c) Preparation of Type 2 smart polymers based on plasma modifications and material binding peptides

This preparation method for Type 2 SP is based on the affinity between specific peptide chains (peptides) and specific Type 1 SP, obtained through treatment with plasma and gas. Each material binding peptide (MBP) has affinity for its own Type 1 SP (Table 3.4), and they can either be synthesized or commercially acquired.

In this way, the peptides can be attached to the biomolecules through chemical conjugation, and in a later step, the modified protein is anchored to the Type 1 SP, given the affinity of MBP for the substrate. The process can also be carried out in the opposite way, that is, anchoring the MBP to the Type 1 SP, and attaching the biomolecule to be immobilized in a later step.

Table 3.4: Combinations of Type 1 SP and material binding peptides (MBP) [162].

Type 1 SP	MBP
Polystyrene treated with plasma and O_2	MBP: PS19-1 and PS19-6
Polymethyl methacrylate treated with plasma and O_2	MBP: c02, PM-OMP25, PMMA-tag
Polysiloxane	MBP: Si-tag
Polycarbonate	MBP: PCOMP6
Poly-L-lactide	MBP: c22

d) Preparation of Type 2 smart polymers with DNA directed immobilization

Small chains of nucleotides (DNA) can be used to attach antibodies to a Type 1 SP. The bond between the small chain of DNA and the antibody must be a covalent bond. Only in that way, the orientation of the immobilized antibodies will be the suitable one (end-on).

e) Preparation of Type 2 smart polymers with enzymes assistance

Sometimes, an enzyme can be used to attach a biomolecule as an antibody. For example, tyrosinase is an enzyme that reacts with tyrosines present in biomolecules surface, turning those tyrosines into *o*-quinones. These compounds react with amine residues ($-NH_2$); therefore, a Type 2 SP with amine residues could directly anchor a tyrosinase-treated biomolecule. This Type 2 SP could be composed of plasma-treated polycarbonate, and a polyallylamine coating.

3.6.2.3 Type 2 *pseudo*-smart polymers

a) Preparation of Type 2 pseudo-smart polymers by entrapment

This type of SP is based on the physical trapping of a biomolecule, preferably an enzyme, in a polymer matrix. It has been included in the "pseudo" category as the Type 1 SP is typically replaced by an electrode (usually platinum). However, the only function of the electrode is to transform the biological signal into an electrical signal (amperometric, potentiometric, etc.). That is, if the response of the system is a visual signal (spectrophotometric), the support could also be a plasma-modified PS.

Ion-exchange polyelectrolytes such as NAFION® are used to make these Type 2 *pseudo*-SP. NAFION® (Figure 3.22) is a commercial copolymer based on tetrafluoroethylene and perfluoroalkyl vinyl ether and it is commonly used in biological applications due to its excellent biocompatibility with biological tissues.

The preparation of these Type 2 *pseudo*-SP is very simple. NAFION® can be commercially purchased as a solution of the polymer in a mixture of alcohols, so a drop of the polymer solution and a drop of the enzyme solution are placed on the

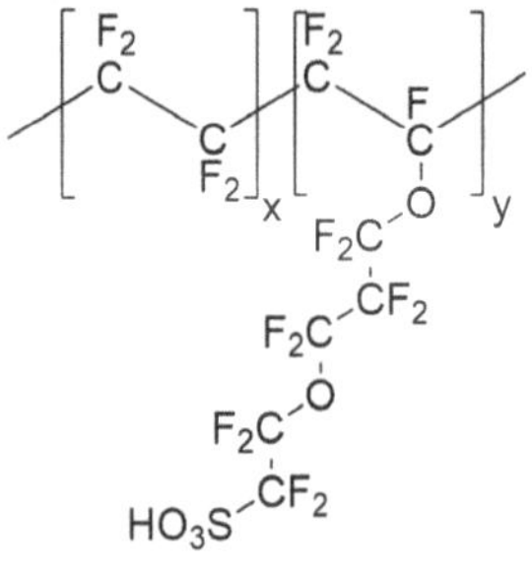

Figure 3.22: Chemical structure of NAFION®.

electrode, and the solvents evaporated. This process can be difficult to optimize, especially the polymer solution concentration. However, once optimized, good results are obtained. Enzyme activity is maintained at 85–95% even several days after Type 2 *pseudo*-SP are prepared, due to the enzyme-polyelectrolyte complexes formed.

b) Preparation of Type 2 pseudo-smart polymers with self-assembled monolayers

Self-assembled monolayers (SAMs) are long, linear-shaped molecules, which contain active head and tail groups separated by a long hydrocarbon chain. The hydrocarbon chain promotes self-assembly when attached to a surface. Additionally, one of the end groups reacts/interacts with the surface of the support material, and the other end group reacts/interacts with the biomolecule [170, 171].

This group has been included as Type 2 *pseudo*-SP, since the support used is not usually a Type 1 SP, but rather a gold support. The gold–thiol interaction is commonly used to prepare these Type 2 *pseudo*-SP with molecules such as 11-mercaptoundecanoic acid. The thiol group of this molecule interacts with the gold surface, leaving free –COOH groups to anchor proteins through the formation of amide bonds.

In a certain way, APTES could be considered a SAM, because it contains functional groups to interact with the support surface at one end –OH), groups for protein attaching at the other end –NH$_2$), and a hydrocarbon chain in the central part of the molecule. Since this chain of hydrocarbons is very small (only three carbons), it has not been considered as a long hydrocarbon chain and thus it has been excluded from this group. Figure 3.23 shows the differences between APTES and 11-mercaptoundecanoic acid in terms of their chemical structure.

c) Preparation of Type 2 pseudo-smart polymers with layer-by-layer materials

LbL systems can be based on Type 1 SP (such as plasma modified PS) but they are typically prepared on supports such as indium tin oxide (ITO) electrodes. LbL

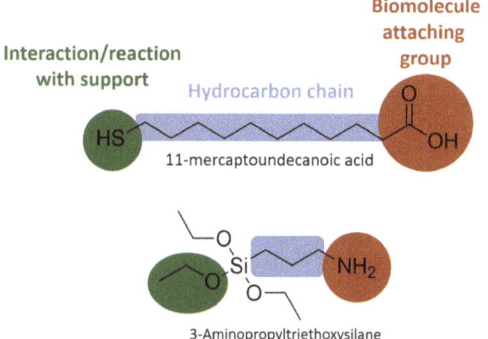

Figure 3.23: Chemical structures of APTES and 11-mercaptoundecanoic acid. The long chain of hydrocarbons of 11-mercaptoundecanoic acid favours self-assembly, while that APTES actually polymerizes through the free –OH groups.

materials are composed of multilayer films sequentially assembled from oppositely charged polyelectrolytes (generally natural-origin polymers). These multilayer materials can be used for multiple biomedical applications, since the initial substrate (PS, ITO, etc.) can be easily removed once the preparation process of the Type 2 pseudo-SP is finished, giving rise to a free-standing material [172].

Since LbL materials are composed of polyelectrolytes, two large groups of polymers are differentiated for the preparation of these materials, polycations and polyanions. Among the most prominent polycations polyethyleneimines stand out, poly (2-(dimethylamino)ethyl methacrylate) or chitosan. On the other hand, poly(vinyl sulphonate) and poly(acrylic acid) are commonly used as polyanions.

In this sense, the immobilized biomolecules are usually enzymes, due to the simplicity and versatility of the method to immobilize these biomolecules without affecting the enzymatic activity. The films are prepared by immersion of the substrate (ITO electrode) in a solution of the polycation and the enzyme, followed by a washing process in the appropriate aqueous buffer (depending on the enzyme), and finally, the immersion of the system in the polyanion solution. This process is repeated as many times as necessary to obtain a workable material or to reach the desired amount of immobilized enzyme.

d) Preparation of Type 2 pseudo-smart polymers with Langmuir–Blodgett films

LB films are monolayer-based nanostructured systems, conceptually similar to SAMs. In fact, the assembled monolayers are composed of amphiphilic molecules containing a polar head (hydrophilic), an apolar tail (hydrophobic) and Langmuir monolayers. LB films can contain one or more monolayers transferred from the liquid–gas interface to the solid supports, during the vertical passage of the support through the monolayers (Figure 3.24) [173].

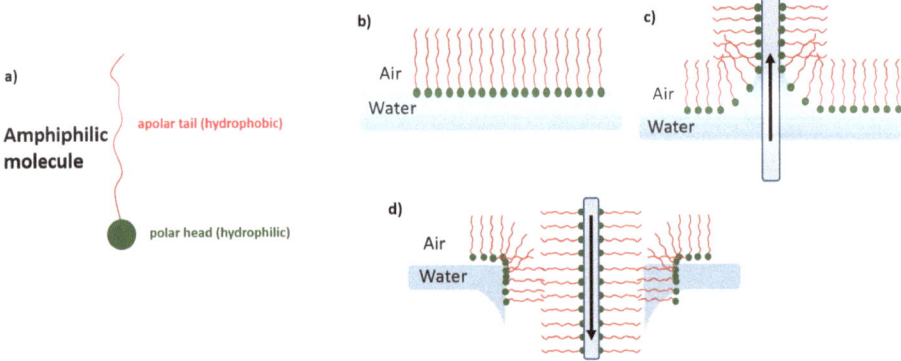

Figure 3.24: Schematic representation of the preparation of Type 2 *pseudo*-smart polymers with Langmuir–Blodgett (LB) films: (a) representation of an amphiphilic molecule; (b) the amphiphilic molecule dissolved in an organic solvent is sprayed onto an aqueous surface, and the molecules are arranged according to the polarity; (c) the support is passed through the solution in a vertical and upward direction, in such a way that the polar heads interact with the support, and the first layer is generated; (d) the support containing the first layer is passed through the solution in a vertical and downward direction, in such a way that the apolar tails interact with each other, and the second layer is formed. This process can be done as many times as needed.

In the first step, the amphiphilic molecule (as stearic acid), the chosen polymer (as poly(3-hexyl thiophene) or similar) and the enzyme are dissolved in a volatile organic solvent such as chloroform. This solution is spread over water, specifically through the air/water interface, in such a way that the organic solvent evaporates and the stearic acid molecules are arranged as shown in Figure 3.24b. As the support material (usually an ITO electrode) passes vertically through the solution, the stearic acid molecules and the poly(3-hexyl thiophene) molecules are arranged as shown in Figure 3.24c, entrapping the enzyme.

There are numerous alternatives to poly(3-hexyl thiophene), such as poly(3-octyl thiophene), or even copolymers based on a thiophene monomer with a polar moiety, and another thiophene monomer with an apolar moiety. This further improves the arrangement of the molecules correctly as shown in Figure 3.25 [174].

3.6.2.4 Type 3 smart polymers
The preparation of Type 3 SP is based on a Type 2 SP, such as any of those previously described. Type 2 SP improve orientation over their previous ones (Type 1 SP), but efforts are still being made to further improve orientation. Type 3 SP are the most advanced type, achieving near-perfect orientation of immobilized biomolecules, especially antibodies. This is achieved by anchoring (at least) one orienting-protein to the

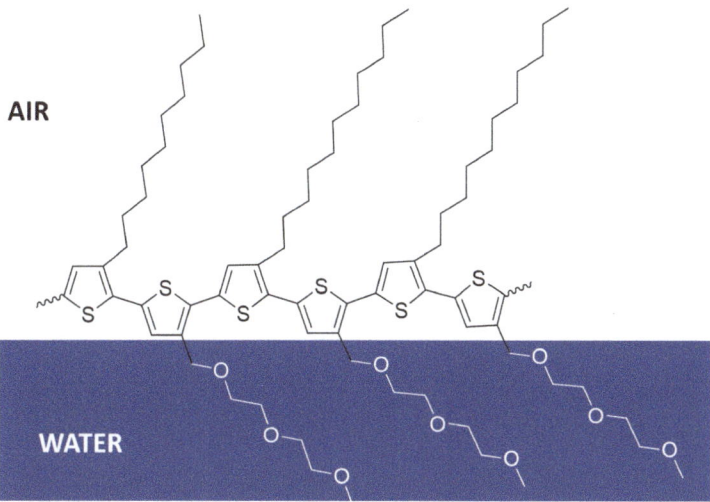

Figure 3.25: Example of Type 2 *pseudo*-smart polymer prepared by Langmuir–Blodgett (LB) procedure. The example combines thiophene monomers both with polar and apolar moieties.

surface of Type 2 SP. Depending on the nature of that orienting-protein, different types of Type 3 SP can be differenced [162].

a) Preparation of Type 3 smart polymers based on Fc binding protein

This type of Type 3 SP is based on anchoring an Fc binding protein (FBP) to a Type 2 SP. The best-known FBPs are protein A (protein originally found in the cell wall of the bacteria *Staphylococcus aureus*, capable of recognizing molecules in extracellular matrix, which is used in biochemical research because of its ability to bind immunoglobulins) or protein G (similar to protein A but with different binding specificities). If an antibody is usually represented by the form "Y", the lower part of that "Y" is known as the crystallizable region, or Fc. The FBPs, conceptually, suppose a perfect attachment for the Fc zone of the antibodies, which implies a perfect orientation of the same (Figure 3.26).

A FBP could be anchored directly on Type 1 or Type 2 SP, but the efficiency of the process is lower since many of the FBPs would not be well placed, and therefore the orientation of the immobilized antibody would not be the ideal one.

b) Preparation of Type 3 smart polymers based on streptavidin/avidin and biotin interaction

Streptavidin and avidin are tetrameric proteins with a high affinity for biotin, a simple and small organic compound (244.3 Da). This feature of biotin is of great importance since this biotinylation process does not affect the conformation of the

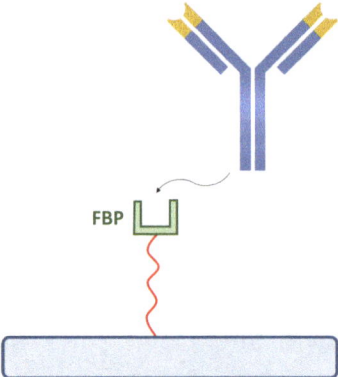

Figure 3.26: Schematic representation of a Type 3 SP structure and the attaching process of an antibody.

biomolecule or its biological activity. This affinity can be used to prepare Type 3 SP following two strategies: Attaching avidin on the surface of Type 2 SP or modifying the material's surface with biotin, although this last strategy is less common.

The attachment of avidin on the surface of Type 2 SP can be carried out through the formation of covalent bonds, with chemical reactions between aldehyde groups or amine groups present on the surface of Type 2 SP. In this way, avidin is immobilized, and a Type 3 SP is obtained. Avidin is usually represented as a cubicle since it has four sites where it can interact with biotin.

On the other hand, the FBP to be attached is biotinylated (Figure 3.27). This process consists of forming an amide bond between the carboxyl group of biotin and the amine side groups present in the FBP. The reaction is carried out in the presence of EDC, and once biotinylated, the FBP is anchored on the surface of the material, generating a Type 3 SP through the avidin/biotin interaction.

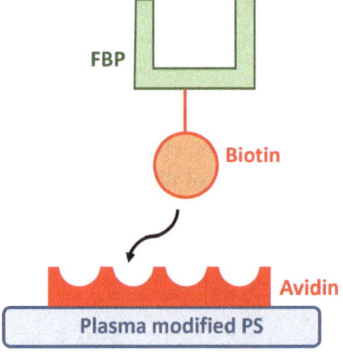

Figure 3.27: Schematic representation of the attaching process of a biotinylated FBP in a Type 2 SP formed by a plasma modified PS and avidin.

4 Emerging applications of smart polymers in biomedicine

4.1 Introduction

Biological and biomedical applications have emerged as the most important research field of smart polymers. Due to their intrinsic characteristics, the number of biological or biomedical applications in which responsive polymers play a central role is spreading quickly. From the early diagnosis of numerous diseases through the detection and immobilization of biomolecules to the selective release of specific drugs at the cellular level, the use of smart polymers has increased greatly in the last years, boosted at the same time for the synthesis and development of new biocompatible and biodegradable polymers. In this part of the book, we will describe three of the main applications of smart polymers in the biological field: smart polymers in drug delivery, new developments in tissue engineering and regenerative medicine, and finally the importance of responsive polymers in cell therapy and precision medicine.

4.2 Drug delivery

4.2.1 Introduction

Since the origin of humanity, civilizations have tried to find the best ways to apply natural remedies to heal wounds and cure diseases, which sometimes involved boiling seeds in water, the preparation of cataplasms for extended performance or mixing with other components such as fats or with other ingredients called excipients to make them more stable or to make their administration easier. Together with the industrial revolution, the production of drugs improved, and the preparation of large-scale medicines available for the population became a reality [175]. Soon, pharmacists and physicians realized that the release rate played a key role in obtaining therapeutic benefits and minimizing the side effects and toxicity, giving rise to a series of generations of controlled releasing systems [176, 177].

The first generation of controlled release systems (*rate-programmed drug release*) consisted of a designed drug dosage form that liberates the drug according to an established rate, minimizing the number of intakes, and without considering the patients' status. Therefore, new excipients were developed around the 1970s as polymer science advanced, controlling the drug release relying on mechanisms related to diffusion, ion exchange or osmosis. These systems allowed the intake of a lower dose of the drug, improving treatments and diminishing side effects in patients [178].

https://doi.org/10.1515/9781501522468-004

In a step forward towards the control of the releasing rate and the releasing site, a new generation of liberating systems was developed (*activation-modulated drug release*). These new systems were born due to the need to protect some drugs from harsh environments in the body, allowing the release in a specific region and minimizing adverse events in other sites. Examples of these systems include polymers with pH or time-dependent solubility or swelling, or suffering enzymatic degradation at the desired body sites [179].

These first and second generation controlled release systems are well established in current therapeutic treatments of different diseases, and polymers have been used as stabilizers or solubilizing agents in the composition of the drug, or even as mechanical support to the release systems [180]. However, current research is directed to the obtention of systems that, in addition to the benefits provided by the first and second generation systems, can respond depending on the evolution of the illness and fit the state of the body: the third generation controlled system, a *feedback-regulated drug release*. This last system needs the presence of active excipients instead of the previous passive ones to control the releasing rate according to the concentration or intensity of a triggering agent as an indicator of the state of the illness, which involves the incorporation of drug molecules in polymer matrices for controlled delivery to tune the pharmacokinetic and biopharmaceutic properties [181, 182]. This way, the excipients act then as sensors and actuators, mimicking the recognition mechanisms that take place in membrane receptors, enzymes or antibodies, as the living systems do when a change in the local environment triggers a change in a property (solubility, charge, size, shape or conformation) [183]. After the drug is released and the concentration of the triggering agent has decreased, the drug release is stopped. This means that these systems interact proactively with the biological functions, receiving information from the body and modifying their behaviour or function, ideally, as a function of the illness progression [184].

The great advance in the generation of these smart biomaterials capable of being in contact with biological substances, cells or tissues without causing damage became possible with the integration of materials science and engineering principles, focusing on the relationship between their processing, structure and their properties [185]. In this sense, polymers play an important part in the evolution of these biomaterials, and soon the confluence of polymer science with biomedicine became unavoidable [186]. Polymers constitute structures that are cheap, stable, versatile, can be prepared with a number of architectures (linear, hyperbranched, cross-linked, hybrid, etc.), with different functions and characteristics, and, as a result, great control of their physicochemical features and functionality is possible [187, 188]. On the other hand, contrary to the large number of scientific papers describing drug delivery materials, some considerations must be taken into account (Figure 4.1), and thus only a few of them have continued to the status of the clinical trial to test their efficiency and safety, cytotoxicity, genotoxicity and antigenicity, not only in their first administration but also after prolonged used. Many of them fail due to limited bioavailability, poor efficiency or other issues [188].

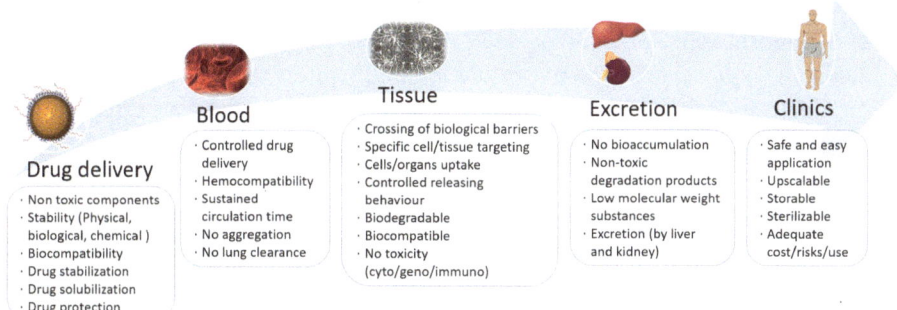

Drug delivery
· Non toxic components
· Stability (Physical, biological, chemical)
· Biocompatibility
· Drug stabilization
· Drug solubilization
· Drug protection

Blood
· Controlled drug delivery
· Hemocompatibility
· Sustained circulation time
· No aggregation
· No lung clearance

Tissue
· Crossing of biological barriers
· Specific cell/tissue targeting
· Cells/organs uptake
· Controlled releasing behaviour
· Biodegradable
· Biocompatible
· No toxicity (cyto/geno/immuno)

Excretion
· No bioaccumulation
· Non-toxic degradation products
· Low molecular weight substances
· Excretion (by liver and kidney)

Clinics
· Safe and easy application
· Upscalable
· Storable
· Sterilizable
· Adequate cost/risks/use

Figure 4.1: Considerations for a smart drug delivery system.

4.2.2 Polymers for drug delivery

Drug delivery systems (DDS) can be especially valuable, for example, in scenarios when the biological medium can degrade the drug before reaching the site of action, when the drug is particularly toxic and the contact with other organs or cells needs to be avoided, when a sustained and controlled drug concentration in the blood is needed, when the drug is intended for very difficult to access cellular structures, when it is intended to follow circadian releasing rhythms or when the patients compliance needs to be improved [189, 190]. On the other hand, these systems also have a series of drawbacks due to their possible dose dumping or toxicity of the materials used, need of surgery to insert or remove the DDS, poor availability, delay in start of action and high preparation costs.

Small molecule drugs sometimes have low stability, poor bioavailability, physical or chemical instability, short solubility, short circulation time or toxicity, and their therapeutic efficiency is limited. Other biodrugs such as peptides, proteins or nucleic acids normally have low stability and are removed from the body very fast. To overcome these obstacles, stimuli-responsive polymers as novel DDSs are being fuelled, and methodologies outside the polymer field are not widely available [191, 192].

The key objective of effective DDS is to improve the pharmacodynamics and pharmacokinetics of drugs to be delivered at the right moment, in the exact place and in the necessary amount. Therefore, the main approaches followed to enable this are [177, 193]:

- *Controlled release*: The efficacy of a drug is enhanced when its concentration is maintained in the therapeutic window. Spatial and temporal release is regulated by therapeutics loaded in polymer carriers, by controlling the speed of dissolution, drug distribution and degradation of the carrier.
- *Targeted delivery*: If the drug is released in the target cell, tissue or organ, the toxicity is minimized and the efficacy is improved. This can be accomplished by covering or conjugating the carrier with functional molecules,

moieties or macromolecules that bind specific polysaccharides, nucleic acids or cell receptor proteins.

– *Solubility improvement*: The encapsulation of lipophilic or hydrophobic drugs in drug delivery carriers or their conjugation with a polymer improve their solubility, enhancing the effectiveness of a therapeutic candidate.

The selection of the polymer for DDS according to their compatibility and desired drug release kinetics is guided both by the characteristics of the DDS and by the type of drug, as summarized in Table 4.1 [192]. The main types of polymeric DDS are polymer drug conjugates, hydrogels and colloidal carriers (micelles, micro or nanoparticles and micro or nanogels). The description of the types of DDS and their fabrication methods is a very extensive field, and their discussion can be found in the works of Mishra *et al.* and Yadav *et al.* [194, 195]. However, a summary of the advantages and weaknesses of each polymer type as DDS are described in Table 4.2 [194, 195].

Table 4.1: Aspects needed for the process of formulating a DDS.

Drug	Drug delivery system	Polymer selection
Drug property (stability, solubility, chemistry, potency) Desired site of action and release rate Delivery challenge (related with the drug)	Route of administration Loading capacity Longevity of release Characteristics: size, shape, flexibility, inclusion of targeting moieties, hydrophobicity	Compatibility with the drug Desired release kinetics

Table 4.2: Advantages and limitations of the different types of drug delivery systems.

DDS polymer type	Advantages	Limitations
Drug conjugates Hydrophilic polymers Dendrimers	Low drug degradation and immunogenicity Increased drug hydrodynamic radius reduces clearance and extends circulation half-life	Low loading capacity Sustained but uncontrolled release Conjugation can cause a decrease in the activity of the drug
Hydrogels Natural polymers Biodegradable polymers Hydrophilic polymers	Broad release lapse of time Useful for targeted delivery Infrequent dosing improves patients approve	Usefulness limited due to drug solubility Large needle or incision needed for delivery Possible local dose dumping

Table 4.2 (continued)

DDS polymer type	Advantages	Limitations
Micelles/ liposomes Amphiphilic block copolymers	Facile synthesis Hydrophobic drugs show enhanced solubility	A bit unstable. Sometimes cross-linking is required
Microparticles Natural polymers Biodegradable polymers	Continuous release A variety of drugs can be encapsulated	Break can cause local toxicity
Nanoparticles Natural polymers Biodegradable polymers	Improved retention and permeation into tumour or tissue due to their small size Enhanced stability	Non-specific uptake in reticular endothelial system

4.2.3 Stimuli-responsive polymers

Smart polymers experience a sudden variation in a physical property as a non-linear answer to a minor environmental change until a critical point is reached, and then they are capable of going back to their initial form after the stimulus is stopped. These changes are reversible, and related to interaction with solvents, hydrophilic/lipophilic rates, physical state, form and solubility and conductivity, causing different responses in the materials, including the formation/destruction of secondary interactions (hydrophobic, van de Waals, hydrogen bonding or electrostatic), simple reactions (acid–base, oxidation/reduction or hydrolysis) or even dramatic conformational changes in the polymeric structure [196, 197].

These types of smart polymers have some advantages such as the versatility in their preparation, sustained release to obtain the desired therapeutic concentration, reduced dosing intakes, reduction of side effects or enhanced stability [196]. They must also possess a number of attributes such as biocompatibility, biodegradability, high drug charging ability and stability, controlled release profile, lack of reproductive or systemic toxicity, carcinogenicity and immunogenicity [192, 196, 197].

Smart polymers for drug delivery can also be classified according to the type of triggering stimuli responsible of their change in: temperature, pH, light, electric/ magnetic field, enzyme or antigen sensitive polymers, polymers sensitive to biochemical signals and dual stimuli-sensitive polymers [198, 199]. Also, the stimuli can be applied externally (thermal, photo, magnetic and electrical stimuli-responsive polymers) or triggered by a physiological stimulus (thermal, pH, enzyme, biological, redox, antigen stimuli-responsive polymers).

Drugs can be encapsulated in the smart polymers in different ways, such as by swelling the dry material in a solution containing the drug until equilibrium, by preparing the gel from a combination of the monomer (or monomers, cross-linkers and initiators) or the polymer and the drug, and also, drugs can be chemically bound to the polymer.

In this chapter, the fundamentals and the mechanism of action of different stimuli-responsive polymers for drug releasing applications are explored, and some specific examples of their applications found in the literature are reviewed and discussed in tables.

4.2.3.1 Temperature sensitive polymers

The most studied responsive polymeric materials for drug delivery are the ones that show a phase transition answer to body temperature variations. Temperature can act both as an internal inducement when the local temperature in the body increases due to a pathological condition (e.g. a tumour or an inflammation), or as an external stimulus, when the temperature is applied from outside. Regarding the latter, many approaches can be applied to increase local temperature to initiate drug delivery, such as ultrasound mediated heating, thermal heating or photoillumination [200, 201].

Typically, temperature-responsive polymers are formed from amphiphilic segments with hydrophilic groups such as secondary, tertiary or quaternary amines, amides or carbonyls, and hydrophobic groups including propyl, ethyl or methyl moieties. These segments can be both monomers or polymer blocks, and inter- and intramolecular interactions between the hydrophobic parts can lead to chain aggregation or cross-linking [198]. These polymers show phase transitions at certain temperatures, triggering a rapid change in their solvated form and their global size as a result of the variations of the polymer–solvent and polymer–polymer interactions. With increasing temperatures, the intensification in the Brownian mobility of the solvent molecules raises the entropy of the system, diminishing the intensity of solvent–polymer interactions. The specific temperatures in this case are called critical solution temperatures (CTS), and thus, temperature sensitive polymers are divided in two categories as mentioned in **Section 2.2** of this book: the ones possessing a lower critical solution temperatures (LCST) and the ones showing an upper critical solution temperature (UCST). The former turns insoluble when heating above the LCTS, undergoing a phase transition, and the latter became soluble upon heating. A general example for LCTS polymers is schematically represented in Figure 4.2a [200] where if the local temperature in the environment surrounding the self-assembled carrier containing the drug is rather superior to the LCTS, polymer chains dehydrate becoming more hydrophobic, and so the system collapses, releasing the encapsulated drug [202].

This fact is explained by the "hydrophobic effect" [38, 203]. The presence of H-bonds among water molecules and polymer hydrophilic moieties at low temperatures results in high solubility. At higher temperatures, the hydrophobic interactions

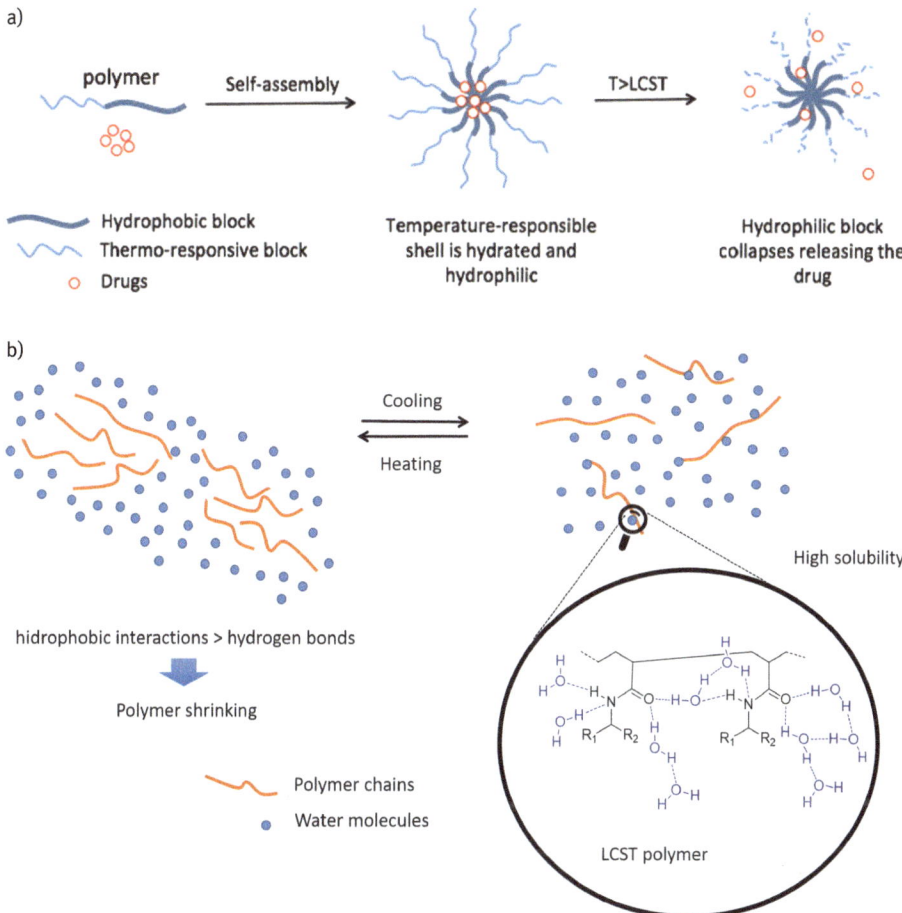

a)

polymer → Self-assembly → T>LCST

- Hydrophobic block
- Thermo-responsive block
- Drugs

Temperature-responsible shell is hydrated and hydrophilic

Hydrophilic block collapses releasing the drug

b)

Cooling ⇌ Heating

High solubility

hidrophobic interactions > hydrogen bonds

Polymer shrinking

- Polymer chains
- Water molecules

LCST polymer

Figure 4.2: (a) Scheme of temperature-responsive amphiphilic polymers for drug deliver; (b) disposition of an LCST polymer chains and water molecules upon heating or cooling.

between the hydrophobic groups prevail over hydrogen bonds, and recoiling/shrinking is observed in the polymer due to the inter-polymer chain interaction (as schematically represented in Figure 4.2b). For LCST polymers, at a certain temperature, this increase in polymer–polymer association or hydrophobic interactions within the polymer chains leads to polymer micellation or precipitation. In this sense, polymers with LCTS behaviour include poly(N,N-diethylacrylamide), poly(N-alkylacrylamides), poly(ethylene glycol) copolymers (PEG), poly(2-ethyl-2-oxazoline) and some other seminatural polymers such as elastin-like polypeptide poly(GVGVP) (elastomeric polypeptide containing G: glycine, P: proline and V: valine). The transition temperatures of these smart polymers are ideally around 37 °C, with some exceptions, and the temperature can be modified including hydrophobic or hydrophilic comonomers, or some additives after polymerization. Polymers showing UCTS are commonly poly(alkylacrylamides) and poly(acrylic acid) (PAAc). Also, it is possible to prepare dual thermo-responsive polymers showing both UCST and LCST using poly (N-isopropylacrylamide (PNIPAAm)) chains in a poly(N-acryloylglycinamide) network [204]. Some examples of polymers that can be used either as homopolymers or copolymers showing LCTS and UCTS and their CTS are described in Figure 4.3.

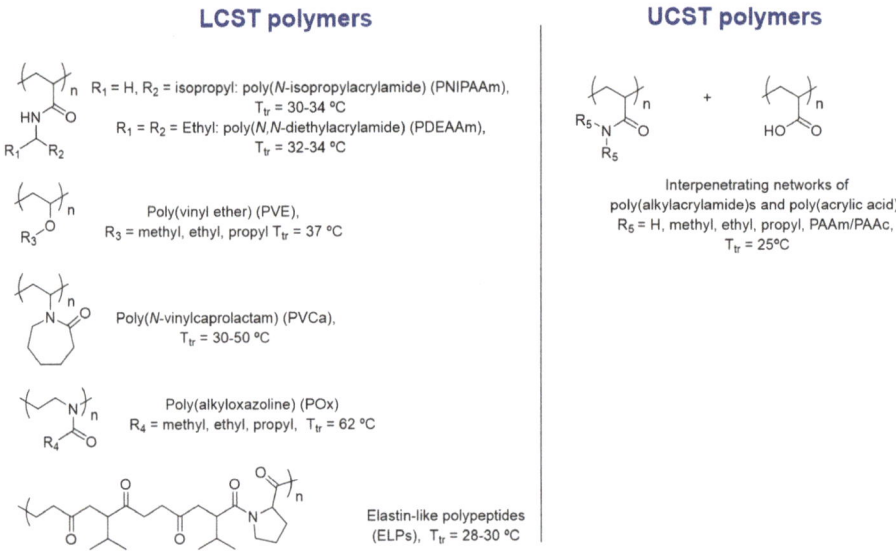

LCST polymers

R_1 = H, R_2 = isopropyl: poly(N-isopropylacrylamide) (PNIPAAm),
T_{tr} = 30-34 °C
R_1 = R_2 = Ethyl: poly(N,N-diethylacrylamide) (PDEAAm),
T_{tr} = 32-34 °C

Poly(vinyl ether) (PVE),
R_3 = methyl, ethyl, propyl T_{tr} = 37 °C

Poly(N-vinylcaprolactam) (PVCa),
T_{tr} = 30-50 °C

Poly(alkyloxazoline) (POx)
R_4 = methyl, ethyl, propyl, T_{tr} = 62 °C

Elastin-like polypeptides
(ELPs), T_{tr} = 28-30 °C

UCST polymers

Interpenetrating networks of
poly(alkylacrylamide)s and poly(acrylic acid),
R_5 = H, methyl, ethyl, propyl, PAAm/PAAc,
T_{tr} = 25°C

Figure 4.3: Chemical structures of common thermo-responsive polymers exhibiting LCST and UCST in water.

The number of different applications of these temperature-sensitive DDS used as hydrogels, liposomes, micelles and drug conjugates is very extensive and diverse, so only some examples of LCTS and UCTS are mentioned in Table 4.3 as a summary.

4.2.3.2 pH-sensitive polymers

pH-responsive polymers are generally obtained through a cycle of deprotonation/ protonation of a weak polyacid or base (polymers containing bases and/or weak acids including amines, phosphoric or carboxylic acids, azo, pyridine or imidazole groups connected to the polymer structure) by a pH-triggered conformation change, which can be reversible or not.

pH-responsive DDS include ionizable polymers with a pKas between 3 and 10. Irreversible pH-responsive polymers lose their structural conformation with a variation of the pH, and include polymers with linkers prone to hydrolysis catalysed by acids or

Table 4.3: Examples of temperature-responsive polymers for DDS.

Type/polymer composition	Drug delivered	System	Discussion	Reference
LCST/poly(N,N-diethylacrylamide)	Insulin	Hydrogels	Increase in the insulin delivery in the first hours compared to poly(N-isopropolylacrylamide) gels	[205]
LCST/poly(2-(2-ethoxyl) ethoxylethyl vinyl ether (EOEOVE)	Doxorubicin (DOX)	Liposomes	Stable liposomes under physiological conditions and above 40 °C release DOX. Tumour growth was suppressed	[206]
LCST/elastin-like polypeptides	DOX	Drug conjugate	The systems are endocytosed by carcinoma cells. Phase transition temperature was at 40 °C, and by combining these polymers with hyperthermia, the cytotoxicity of the system was increased	[207]
LCST/poly(2-oxazolines)	Paclitaxel, Amphotericin B, Cyclosporin A	Micelles	They form micelles as a response to temperature. They show high loading capacity of hydrophobic drugs without loss of activity	[208]
LCST/cytosine-PPG	DOX	Hydrogel	DOX charged nanogels display lower IC50 than free DOX cancer when exposed to temperatures of 40 °C. At 25 °C, nanogels are not cytotoxic	[209]
UCST/poly (acrylamide)/poly (acrylic acid) interpenetrating network	Ibuprofen	Hydrogel	Drug was released faster at 37 °C than at 25 °C	[210]
UCST/poly (acrylamide-co-acrylonitrile)	DOX	Micelles/functionalized nanoparticles	When the transition temperature is reached, the drug is almost instantaneously released. Improved antitumour efficiency	[211, 212]

bases (a schematic representation of two different possibilities of DDSs containing acid responsive bonds and encapsulated drugs is exemplified in Figure 4.4a). On the other hand, reversible pH sensitive polymers are commonly polyelectrolytes with various weak bases or acid functional groups. These polyelectrolytes can donate or accept protons as a function of the environmental pH, causing an alteration of the polymer solubility, size, conformation and so on and thus the drug is released (a schematic representation of a general situation is depicted in Figure 4.4b).

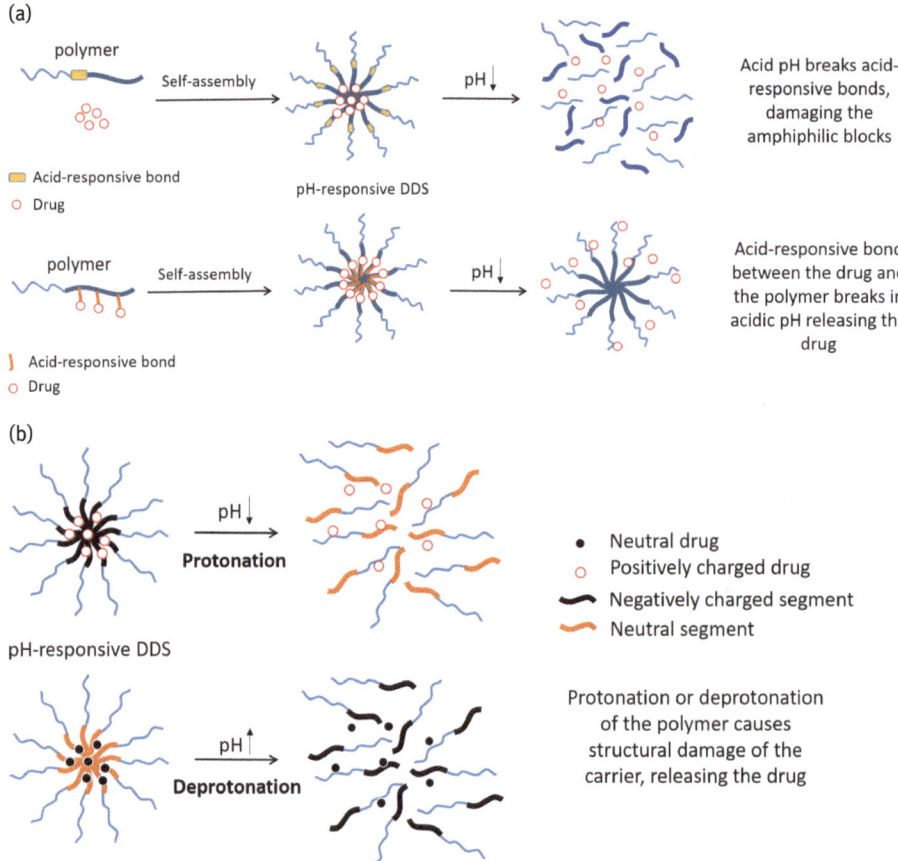

Figure 4.4: pH-responsive mechanisms for drug delivery caused by a change in the pH: (a) break of acid-responsive bonds; (b) protonation/deprotonation of polymers.

The pH response can be tuned to a precise pH according to the pKa of the chosen functional groups. Some examples of structures used for pH sensitive polymers (also natural polymers) in drug delivery uses including basic, acid and polymers containing pH labile bonds are shown in Figure 4.5. There are two main kinds of pH-responsive polymers: polymers with acid or basic moieties. Also, polymers with

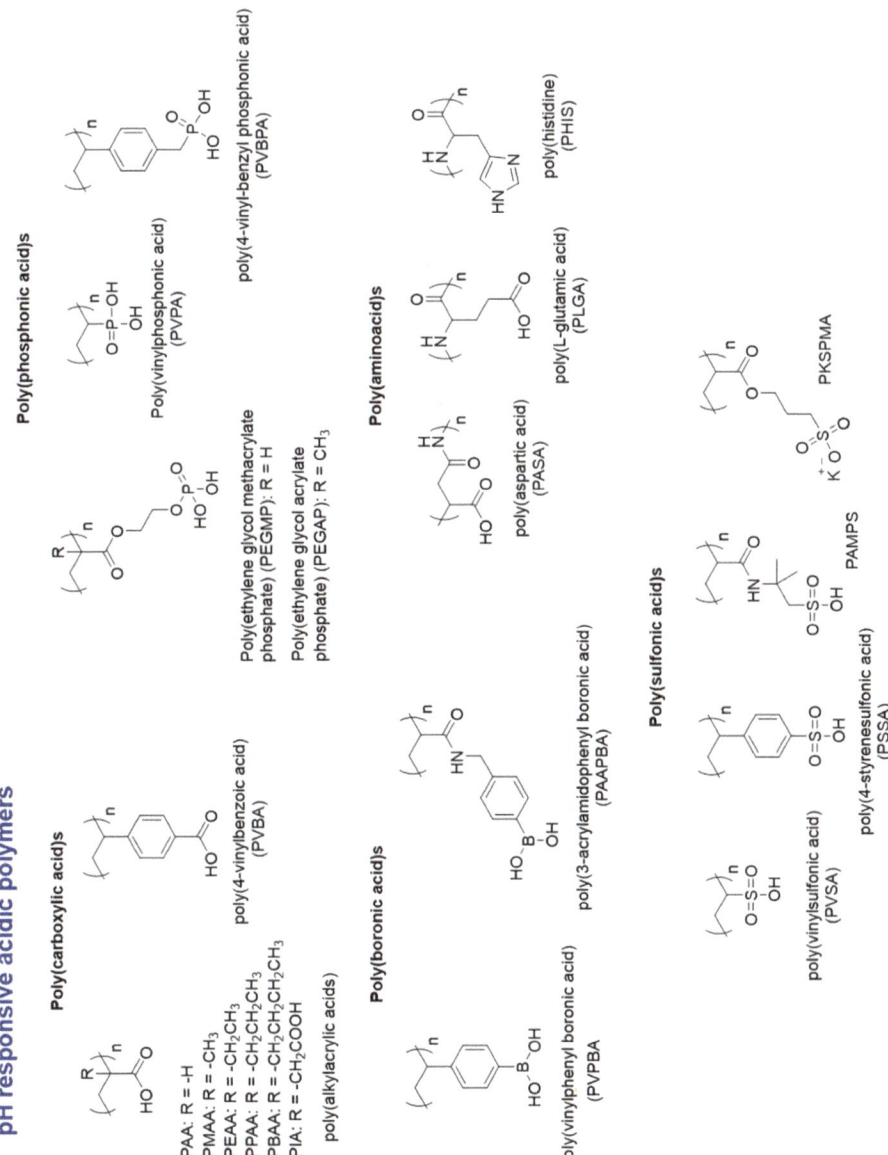

Figure 4.5: Common pH sensitive polymers used in drug delivery applications [23].

pH responsive basic polymers

Polymers containing tertiary or secondary amine groups

poly[(dialkylamino)ethylacrylates]

PDMA: R$_1$ = -CH$_3$, R$_2$ = R$_3$= -CH$_3$
PDEA: R$_1$ = -CH$_3$, R$_2$ = R$_3$= -CH$_2$CH$_3$
PDPAEMA: R$_1$ = -CH$_3$, R$_2$ = R$_3$= -CH$_2$CH$_2$CH$_3$
PDPA: R$_1$ = -CH$_3$, R$_2$ = R$_3$= -CH(CH$_3$)$_2$
PDMAEA: R$_1$ = -H, R$_2$ = R$_3$= -CH$_3$
PtBAEMA: R$_1$ = -CH$_3$, R$_2$ = H, R$_3$= -CH(CH$_3$)$_3$

poly(N,N-dialkylvinylbenzylamine)s

PDMVBA: R$_1$ = R$_2$ = -CH$_3$
PDEVBA: R$_1$ = R$_2$ = -CH$_2$CH$_3$
PDPVBA: R$_1$ = R$_2$ = -CH$_2$CH$_2$CH$_3$

poly[(2-diethylamino)ethylacrylamide]
(PDEAm)

Polymers containing morpholino, pyrrolidine and piperazine groups

R$_1$ =

poly(N-alkyl)ethylmethacrylates

poly[(2-N-morpholine)ethylmethacrylate] (PMEMA)
poly[(N-ethylpyrrolidine)methacrylate] (PEPyMA)

poly[(2-N-morpholine)ethylmethacrylamide]
(PMEMAm)

R$_1$ =

R$_2$ = -CH$_3$
-CH$_2$CH$_3$
-CH$_2$CH$_2$CH$_3$

poly(acryloylmopholine) (PAM)
poly(N-acryloyl-N'-alkenylpiperazine)

Polymers containing pyridine and imidazole groups

R$_1$ =

poly(4-vinylpyridine) (P4VP)
poly(2-vinylpyridine) (P2VP)
poly(N-vinylimidazole) (PVI)

poly[6-(1H-imidazol-1-yl)hexyl-methacrylate]
(PImHeMA)

Dendrimers

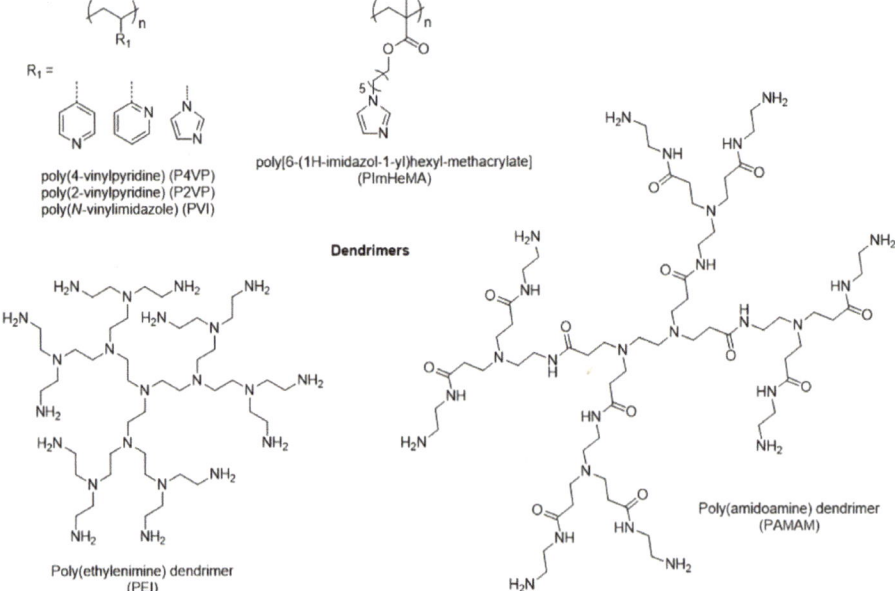

Poly(ethylenimine) dendrimer
(PEI)

Poly(amidoamine) dendrimer
(PAMAM)

Figure 4.5 (continued)

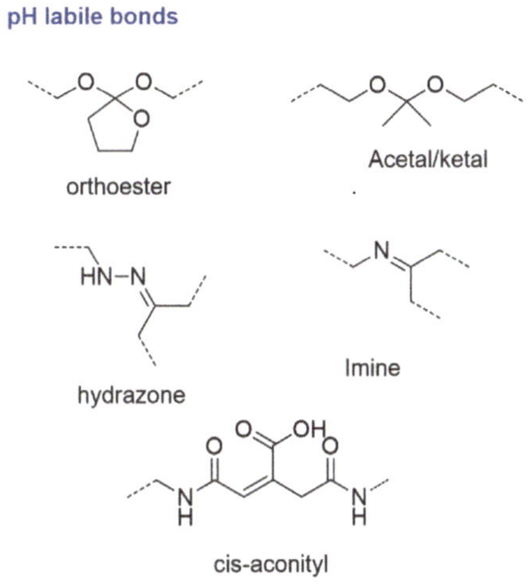

pH labile bonds

Figure 4.5 (continued)

pH labile bonds can be used for DDS design. Weak polyacids accept protons at low pH and donate them at neutral and high pH, while weak polybases accept protons at low pH, and gaining polyelectrolyte nature. The polymer–water interactions through electrostatic or hydrogen bonding in the ionized state result in a greater quantity of polymer associated water of hydration. Also, polymer chain expansion can be observed as a result of the electrostatic repulsion between functional moieties with the same charge. In contrast, polymers in the unionized state show weak inter- and intrachain hydrophobic and dipole–dipole interactions, and polymer aggregation or precipitation can occur through the increased solute interactions and the resulting solvent expulsion [191, 196, 198].

The fact that different pH environments cause a sharp change in physiochemical properties show a great potential in DDS since some tumour tissues have a slightly lower pH than normal tissue. In this sense, a more extensive discussion of the uses of pH-responsive polymers for drug delivery in cancer treatment is found in **Section 4.4** of this book concerning cell therapy. They are usually designed according to the physiologically relevant pH changes, and are typically used in oral controlled release formulation for the gastrointestinal (GI) tract pH variations (GI fluid pH rises gradually from the stomach to the colon) [197, 201]. pH-responsive materials can be designed to meet special features such as the flexibility, mechanical strength, solubility or the degradation rate, being prepared in different forms and shapes to be used in different administration methods such as intraosseous, subcutaneous, intravenous, transdermal or ingestion. As in the case of temperature-sensitive polymers, there are many examples of pH-sensitive applications for DDS, so a discussion of some a few examples of their applications in DDS systems is summarized in Table 4.4. However, these systems need the local pH to vary according to the severity of the illness or the closeness to the unhealthy tissue. In addition, the conservation of the structure can be complex since the system can also be activated during administration.

Table 4.4: Examples of drug delivery uses of pH-responsive polymers.

Polymer composition	Drug delivered	System	Discussion	Reference
N-carboxyethyl chitosan/ dibenzaldehyde-terminated poly (ethylene glycol)	DOX	Hydrogel	pH modification causes physical and chemical variations swelling the system and releasing the drug. Used in human hepatocellular liver carcinoma (HepG2)	[213]
Poly(lactic acid)-poly (ethyleneimine)	DOX	Nanoparticle	Fast release of DOX as pH changes from 7.4 to 5.4. Suppressing proliferation of MCF-7 cells	[214]
Poly(lactic-co-glycolic acid) (PLGA)	Vancomycin	Microsphere	Structural change produces drug delivery. Used to treat osteomyelitis	[215]
Poly(acrylamide)-g-carrageenan and sodium alginate	Ketoprofen	Hydrogel	Drug release is raised when the pH changes from acid to basic. Colon-targeted delivery	[216]
Hyaluronic acid-hidrazone-DOX	DOX	Nanoparticle	Drug liberation was induced by pH gradient. The system showed higher toxicity on tumour cells (HeLa) than in other cells	[217]

4.2.3.3 Redox-potential sensitive polymers

Redox reactions involve an electron transfer between chemical species, leading to the development of covalent bonds and breaking other existing ones. This fact can be used to develop DDSs in specific body sites, since the redox state of the different parts of the body or different cell sites is different. The difference in the redox potential between the extracellular (oxidizing) and the intracellular (reducing) spaces can be used as possible stimulus for the controlled DDSs. Also, an oxidative environment is related with inflammation process, caused by pathological processes such as cancer or arthritis.

This means that if a redox-responsive chemical moiety is somehow linked to a drug, it will remain stable when the redox state is neutral. Choosing a linker sensitive to a certain redox state in the body tissue can then control the drug delivery [28, 191]. For this purpose, the most commonly exploited biological redox couple to trigger the drug release is the glutathione (glutamyl-cysteinyl-glycine; GSH)/glutathione disulphide (GSSG). The presence of the cysteine residue causes GSH to be oxidized by free radicals and reactive oxygen/nitrogen residues and turned into GSSG. Then, GSSG is eliminated from the cells. The ratio GSH/GSSG serves to specify the cellular redox state, which is around 10 in usual environments, but rises up until 1,000 in cancer cells, although the level of GSH also depends on other redox couples (NADH/NAD+, NADPH/NADP+) [218, 219], so this fact can be used to prepare redox-responsive intracellular drug delivery. Common redox sensitive polymers usually include an S–S bond, and some examples are depicted in Figure 4.6.

The preparation of these drug delivery polymeric systems depends on the chemistry of the redox-sensitive motifs, that is the disulphide bonds, showing greater stability in oxidizing extracellular environments, and tends to be reduced to generate thiol groups in reducing environments. These disulphide bonds can be located both in the polymer backbone and in the side chain, being the latter easier to be modified. The number of disulphide bonds can be controlled, so when facing the reducing media, the system with higher proportion of disulphide bonds will have a higher response to fully release the drug [220]. This way, the polymeric carrier will be broken apart in the case of the high amounts of GSH present inside the tumour cell (GSH amount is lower in healthy than in tumour tissue), delivering the active drug (Figure 4.7).

A more extensive discussion of uses of redox-responsive polymers for drug delivery in cancer treatment can be found in **Section 4.4** of this book. However, some examples of micelle and nanoparticles redox-sensitive DDS applications for drug delivery are summarized in Table 4.5, together with their discussion.

4.2.3.4 Light-sensitive polymers

Another smart option for the controlled delivery of drugs are the light-responsive materials. One of the main advantages of these materials is that they can be stimulated non-invasively with a great precision in terms of location and time, reducing the affection to other tissues and systemic toxicity [225, 226]. Current available equipment can apply radiation of specific wavelengths between 2,500 and 380 nm, including ultraviolet (UV), visible or infrared light and lasers of various wavelengths. Blue or UV light is commonly used as activation releasing agent for topical treatments including mucous, skin or eyes. However, UV light has reduced tissue permeation and has harmful effects on healthy tissue [227]. On the other hand, higher wavelength light (infrared) penetrates to deeper tissues; it is innocuous and does not produce as much heating in the application area. Nevertheless, NIR-light sensitive systems show lower efficacy and thus need an extended exposure time to generate beneficial effects, causing damage to the surroundings. Despite the many advantages of these systems, light-responsive polymers are sometimes difficult to synthesize, limiting their bulk production.

Light-sensitive polymers contain functional groups that change their conformation when irradiated. The light-response characteristic is commonly introduced in

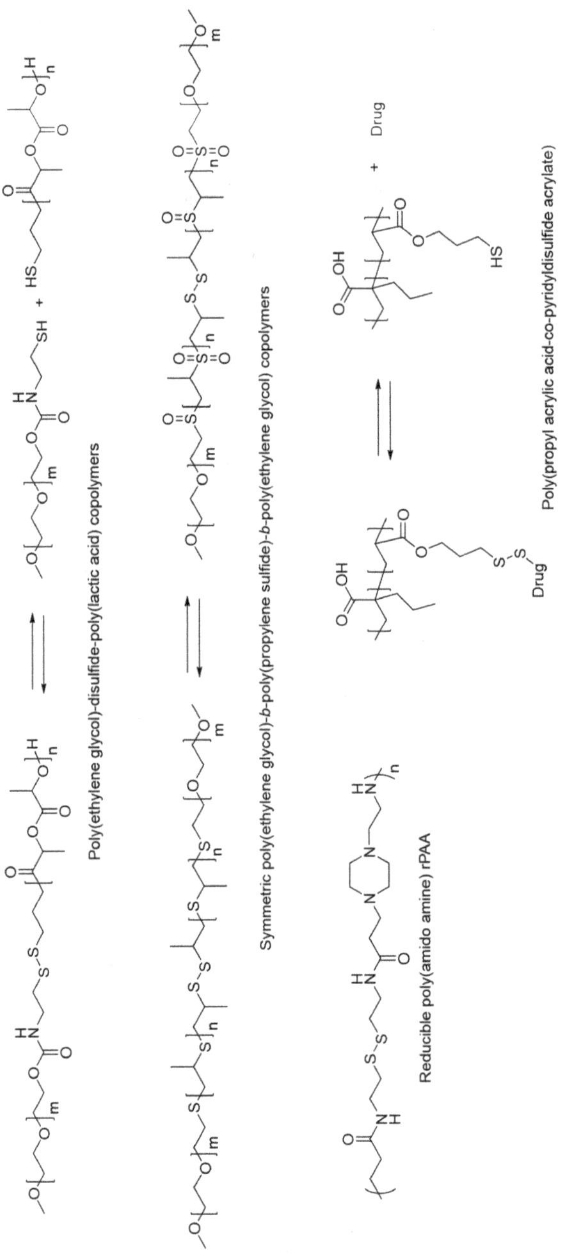

Figure 4.6: Redox-responsive polymers for drug delivery.

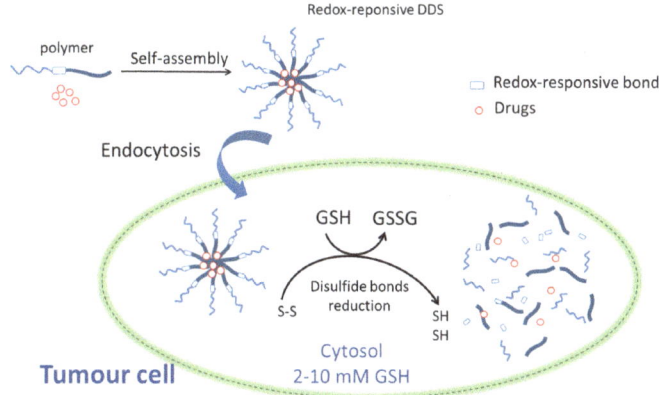

Figure 4.7: Scheme of redox-responsive micelles for drug delivery inside cytosol due to GSH-triggered disassembly. Glutathione (GSH) reduces the disulphide bonds after approaching the cytosol.

Table 4.5: Examples of drug delivery uses of redox-sensitive polymers.

Polymer composition	Drug delivered	System	Discussion	Reference
Poly(ethylene glycol)-b-poly (propylene sulphide)-b-poly (ethyleneimine)	Plasmid DNA	Micelle	DDSs effectively transfected melanoma cells, and tumour development was diminished	[221]
Poly(ethylene glycol)-SS-poly (lactic acid)	Paclitaxel (PTX)	Nanoparticles	PTX delivery from particles was reactive towards GSH. Uniform intracellular distribution	[222]
Hyperbranched poly (2-((2-(acryloyloxy)ethyl) disulphanyl)ethyl 4-cyano -4-(((propylthio) carbonothioyl)-thio)- pentanoate-co-poly(ethylene glycol) methacrylate) (HPAEG)	DOX	Nanoparticles	When the HPAEG-AS1411 nanoparticles charged with DOX enter tumour cells, the disulphide bonds were broken by GSH, liberating the drug	[223]
Poly(ethylene glycol)-block- poly(2-(methacryloyloxy)ethyl 5-(1,2-dithiolan-3-yl) pentanoate) diblock copolymers (PEG-b-PLAHEMA)	DOX	Micelle	Decrease in IC50 of 231 R cancer cells by DOX charged CCMs	[224]

the polymer through a linker that can be broken when exposed to light irradiation of a defined wavelength, such as the *o*-nitrobenzyl light-cleavable linkage [228]. Also, light has been used as a stimulus to disrupt the carrier with a chemical switch such as azobenzene, nitrobenzyl, pyrene, coumarin and spirobenzopyran (Figure 4.8) [229–231]. In this sense, isomerization, dimerization and others of these compounds after irradiation with light induces a change in their polarity or hydrophilicity [232, 233]. A third type of light-sensitive DDSs are based on a polymer containing a photosensitive chromophore. When a certain wavelength is applied, the chromophore absorbs light, and then it is dissipated as heat by a radiationless transition, causing the local temperature to rise (as a function of the chromophore concentration and the intensity of the light), resulting in a variation of the swelling performance of the thermosensitive system [196].

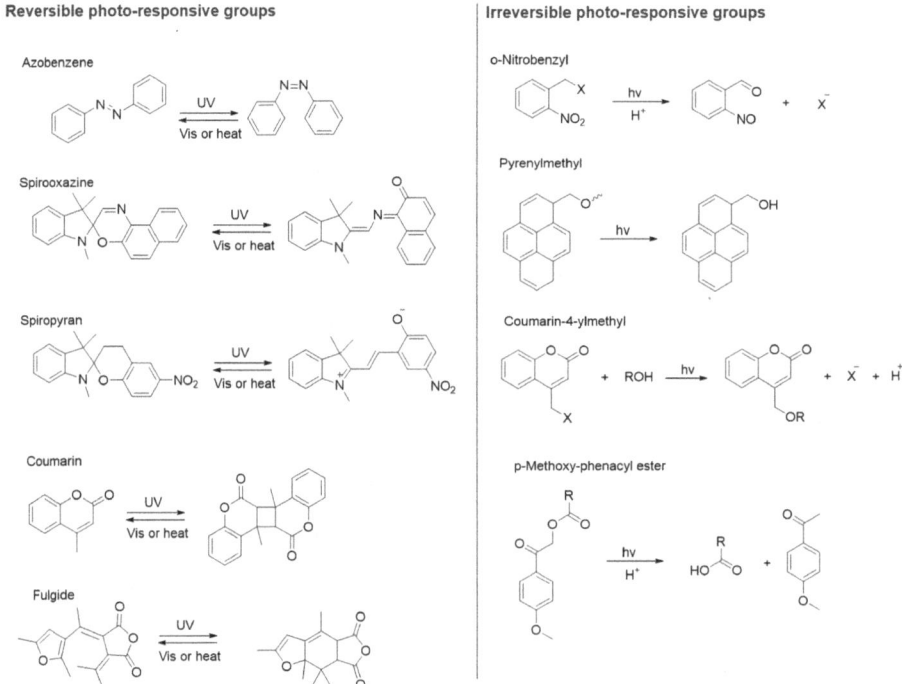

Figure 4.8: Examples of various types of reversible and irreversible light-responsive groups.

Some light-responsive DDSs are intended for a single use, with an irreversible structural transformation in the system, liberating the complete dose, and others can undergo reversible conformational variations when cycles of light/dark are applied, discharging the therapeutic chemical in a pulsatile manner [229].

Figure 4.9 illustrates in a schematic way the general mechanism of different scenarios for photosensitive systems for drug delivery. Light exposure prompts solubility variations in the block modified with the photochromic moiety and the DDS breaks apart.

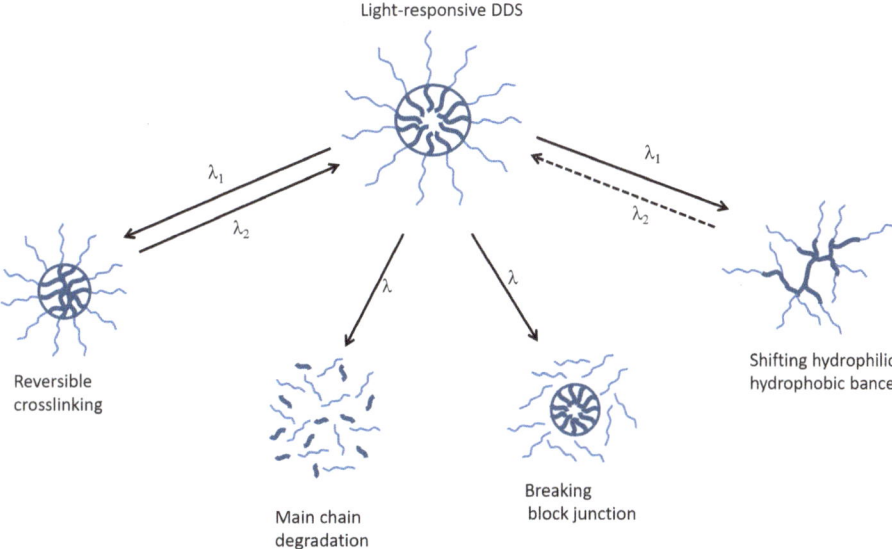

Figure 4.9: Representation of various forms of light-responsive block copolymer micelles. Light causes solubility changes and micelles disassemble.

Once again, the number of examples of different DDS (mostly micelles and nanoparticles) that can be found in the literature with the above-mentioned photoresponsive motifs in their structure is very extensive. Most of the examples found are intended for delivering doxorubicin (DOX) for cancer treatment. Some examples are summarized in Table 4.6.

Table 4.6: Examples of drug delivery uses for photo-responsive polymers.

Polymer composition	Drug delivered	System	Discussion	Reference
Poly(4,5-dimethoxy-2-nitrobenzyl methacrylate)	DOX	Nanoparticle	Upon 980 nm light irradiation, the nanoparticle absorbed the UV from upconversion nanoparticles causing break of the *o*-nitrobenzyl residues, changing the hydrophilic/hydrophobic ratio, disassembling the nanoparticle and releasing the drug	[234]

Table 4.6 (continued)

Polymer composition	Drug delivered	System	Discussion	Reference
Spiropyran-hyperbranched glycerol	Hydrophobic drugs	Micelle	Micelle assembly/disassembly stimulated by UV irradiation to control the drug discharge	[235]
Azobenzene-β-galactose	Hydrophobic drugs	Micelle	Low cytotoxicity. The azobenzene motifs isomerized fast to the polar *cis* isomers with UV light. Used for melanoma A375 cells	[236]
ABA-type triblock copolymer poly(ethylene glycol)-b-poly (ethanedithiol-alt-nitrobenzyl)-b-poly (ethylene glycol)	DOX	Micelles	The carrier is prepared to have light-sensitive *o*-nitrobenzyl bonds so when light is applied, a rapid burst release is induced. Quicker drug discharge in A549 cells after UV irradiation and better anticancer activity	[228]
Spiropyran-containing amphiphilic copolymer, poly(RBM-MAPEG-SPMA)	DOX	Nanoparticle	The DDS varies its hydrophilic–hydrophobic balance turning into hydrophilic when UV light is applied and breaking away from the hydrophobic surface releasing the pre-loaded anticancer drug	[237]

4.2.3.5 Electric field-sensitive polymers

Electric field-responsive polymers modify their physical characteristics due to a small variation in the electric current. Thus, electric stimuli can be generated with devices for transdermal delivery, enabling a defined control through regulating the current intensity, the length of the pulses and the time interval between the electric pulses [238, 239]. These polymers are made using polyelectrolytes with large number of ionizable motifs in the polymer backbone, and thus they have a similar releasing mechanism to those with a pH response [240].

They are processed in the shape of sheets, microparticles or, for subcutaneous implantation, as *in situ* gelling injectable systems. An electric field is applied through the electro-conducting system located on the skin above the polyelectrolyte structure. The ions movement caused by the potential leads to local changes in pH, originating a shrinking or a swelling in the cross-linked polyelectrolytes, leading to drug discharge due to the degradation or bending or the system structure. The explanation of the shrinking in volume because of the disruption of hydrogen interactions between the polymer chains is simultaneously attributed to two different facts. In the first place, to the migration of hydrated H^+ towards the cathode and water loss happening at the anode. In the second place, the electrostatic interaction of negatively charged functional moieties towards the anode results in the formation of stress along the polymeric system backbone [191]. Thus, control of the releasing rate can be obtained by applying pulses of electricity, alternating swelling and shrinking of the system, which has particularly been tested for insulin delivery [241].

The use of intrinsically conducting polymers (ICP) like polypyrrole (PPy), polyaniline (PAni) and poly(3,4-ethylenedioxythiopehene) (PEDOT) (the structure of these materials can be found in Figure 3.5 in Chapter 3) is an alternative to the use of polyelectrolytes. These polymers show electrical conductivity due to the alternating single and double bonds that create an ordered and uninterrupted π-conjugated backbone [242, 243]. Also, in order to control the redox state, ionic dopants are included in the polymerization step, and they can work as ion carriers upon polymer charge variations if the drug is bonded to the carrier. When a potential is applied to these ICPs, reversible oxidation/reduction occurs, altering the polymer change and/or inducing a conformational variation. Reduction or oxidation of the polymer can directly discharge the drug load, increasing the elution as a response to the stimuli (e.g.; in anionic ibuprofen) [244].

Alternatively, the dopant can carry the drug out of the polymer matrix when stimulated, or the polymer can suffer conformational alterations increasing the diffusion [245]. ICPs typically show poorer mechanical properties and no biodegradability, so they are commonly incorporated into natural hydrogels.

Electric stimulation is an easy and economical method to modulate drug delivery, providing reversible and reproducible responses. Examples of applications of these systems for drug delivery are discussed in Table 4.7. The main drawback of these DDSs is that their use is restricted to topical or subdermal implants, normally as hydrogels, since it is necessary to place electrodes in the polymer matrix, and so their applications are consequent with this fact.

4.2.3.6 Magnetic field-sensitive polymers

Magnetic drug carriers can act in three different ways: visualizing the drug carrier in de body by magnetic resonance imaging, controlling the tissue distribution applying an outer magnetic field and triggering drug discharge as a consequence of a local temperature rise by application of an alternating magnetic field [249, 250].

Magnetic field-responsive polymers are generally prepared using magnetic microbeads or Fe_3O_4 superparamagnetic iron oxide nanoparticles (SPIONs), which react to an applied magnetic field. The drug discharge occurs according to the strength of the magnetic field, stopping the drug liberation when the magnetic field is removed [197]. SPION stimulation with an alternating current magnetic field prompts Neel

Table 4.7: Examples of drug delivery uses for electric field-sensitive polymers.

Polymer composition	Drug delivered	System	Discussion	Reference
Agarose/alginate-aniline tetramer	Dexamethasone	Hydrogel	Research in neuroregenerative medicine. Improved hydrogel biocompatibility with neural cells	[246]
Poly(3,4-ethylenedioxypyrrole)	Ibuprofen	Film	Electric stimulation causes a rapid and controlled discharge of ionically attached ibuprofen but not physically trapped ibuprofen	[247]
Polypyrrole	DOX	Nanowires	Efficient system for controlled drug discharge as a function of the applied electric field. Enhanced therapeutic results	[245]
Poly (dimethylaminopropylacrylamide) (PDMAPAA)	Insulin	Hydrogel	Administered subcutaneously in rats. Responding to a stimulus, the drug is released and the glucose levels decrease	[241]
Sodium alginate and carbopol	Diclofenac sodium (DS), diclofenac potassium (DP), and diclofenac diethylammonium (DD)	Hydrogel	As a consequence of a pulsatile electric current, the drug release varies	[248]

relaxation, Brownian losses, or both, generating heat, which can be used to induce conformational changes, break heat-labile covalent bonds, and so on, to intensify drug distribution from the DDS. Temperature increase can be controlled by the time of application and the frequency of oscillation of the magnetic field. A slight rise in the temperature leads to a reversible drug release according to a reversible conformational change in the polymer, and a strong increase results in the polymer carrier rupture, leading to the whole dose drug release [251].

Commonly, these iron oxide nanoparticles are just physically entrapped inside a polymer matrix together with the drug. However, magnetically responsive polymers *per se* can also be prepared by covalent conjugation of ferromagnetic particles, although they are not a common platform for magnetically responsive drug delivery and are not so well explored. Some examples of covalent conjugated magnetic particles other than SPIONS are paramagnetic chelates of gadolinium (Gd) [252]. They are employed as magnetic resonance contrast media for imaging, and as a result of their low molecular weight and the need of high doses, they can be toxic to the human body. In this sense, magnetic chelates of Gd or ferrous and ferric nanoparticles can be functionalized and covalently linked to dendritic polymers to be used in cancer treatment, to target and carry the DDS and fluorophores to the tumour tissue [253, 254]. Some examples of these systems for drug delivery found in the literature are discussed in Table 4.8. This time, the common DDS are nanoparticles and dendrimers and are intended for delivering diverse drugs.

Table 4.8: Examples of drug delivery applications for magnetic field-responsive polymers.

Polymer composition	Drug delivered	System	Discussion	Reference
Polyethylene glycol with azo drug linker	DOX	Nanoparticle	Iron oxide nanoparticles produce heat under an alternating magnetic field, and DOX with thermo-labile azo molecule is released	[255]
Poly (amidoamine) (PAMAM)	Antisense survivin oligodeoxynucleotide	Dendrimer	The drug was efficiently internalized in the cell, inhibiting its growth as a result of dose and time variations	[254]
Polyethylenimine (PEI)	siRNA	Nanoparticle	PEI-SPIOs were able to silence specific Genes *in vitro* without being cytotoxic and damaging siRNA stability	[256]
Poly (amidoamine)	Avidin	Dendrimer	Visible changes in ovarian tumours were found after an efficient delivery of chelated Gd(III)-fluorophores by the drug delivery system	[253]

4.2.3.7 Bioresponsive polymers

Bioresponsive polymers are classified into glucose-sensitive polymers, enzyme-responsive polymers and antigen-responsive polymers. These DDS are gaining relevance in various biomedical applications, as they react to the stimuli intrinsically existing in the body. They have in common that the response comes from common functional groups known to interact with biologically significant substances. Biomolecules have higher specificity than those responding to chemical or physical stimuli.

a) Glucose-sensitive polymers

The biochemical signal in this case is the rise in the blood sugar level, and it is used to control insulin release in diabetes therapy. They mimic endogenous insulin secretion to transport blood sugar to the cells. Their main weaknesses are the possible non-biocompatibility and their short response time. There are three main glucose-sensitive polymeric systems: built on glucose-boronic acid complexation in polymeric chains, based on the interaction of lectin (Concanavalin A (Con A)) with glucose, and based on the enzyme glucose oxidase, which turns glucose to gluconic acid [257]. Figure 4.10 shows the main mechanisms and structures used in these systems.

- *Glucose-boronic acid complexation:* Polymers having phenylboronic acid moieties form a complex with the hydroxyl groups of sugars [258]. Diol concentration or environmental pH controls the water solubility of boronic acid moieties. Polymers with neutral phenyl boronic acids are hydrophobic, and anionic boronate ester groups are hydrophilic. When glucose concentration increases in the human body, the ratio of anionic boronate/native boronic form increases, turning the hydrophobic system into hydrophilic, thus releasing insulin [191].
- *Lectin (protein)-based glucose sensitive polymers:* Lectins are multivalent proteins present at the cell's surface with specific glucose and mannose binging ability. A lectin commonly used in insulin-modulated drug delivery is Con A, with four binding sites, and it is useful for immobilizing chemically modified insulin (glycosylated insulin) [259]. In glucose-free medium, no release takes place. When glucose is present, a competitive mechanism occurs. When free glucose reaches a certain concentration, glycosylated insulin is easily displaced from the binding to Con A and so discharged [196].
- *Glucose-oxidase (GOx)-based glucose-sensitive polymers:* The oxidation of glucose is catalysed by GOx, forming H_2O_2 while inducing hypoxia, and yielding glucono-1,5-lactone. This product suffers hydrolysis when water is present to yield gluconic acid, modifying the local pH. Glucose-sensitive polymers based on GOx respond to these by-products due to the presence of chemically or physically sensitive functional groups which are embedded or linked to the polymer. Generally, these systems are prepared from pH sensitive polymers. Polymers used for this type of DDS can be prepared from chitosan, or are prepared from monomers such as dimethylaminoethyl, 2-hydroxyethyl or 3-trimethoxysilypropyl methacrylate. As blood

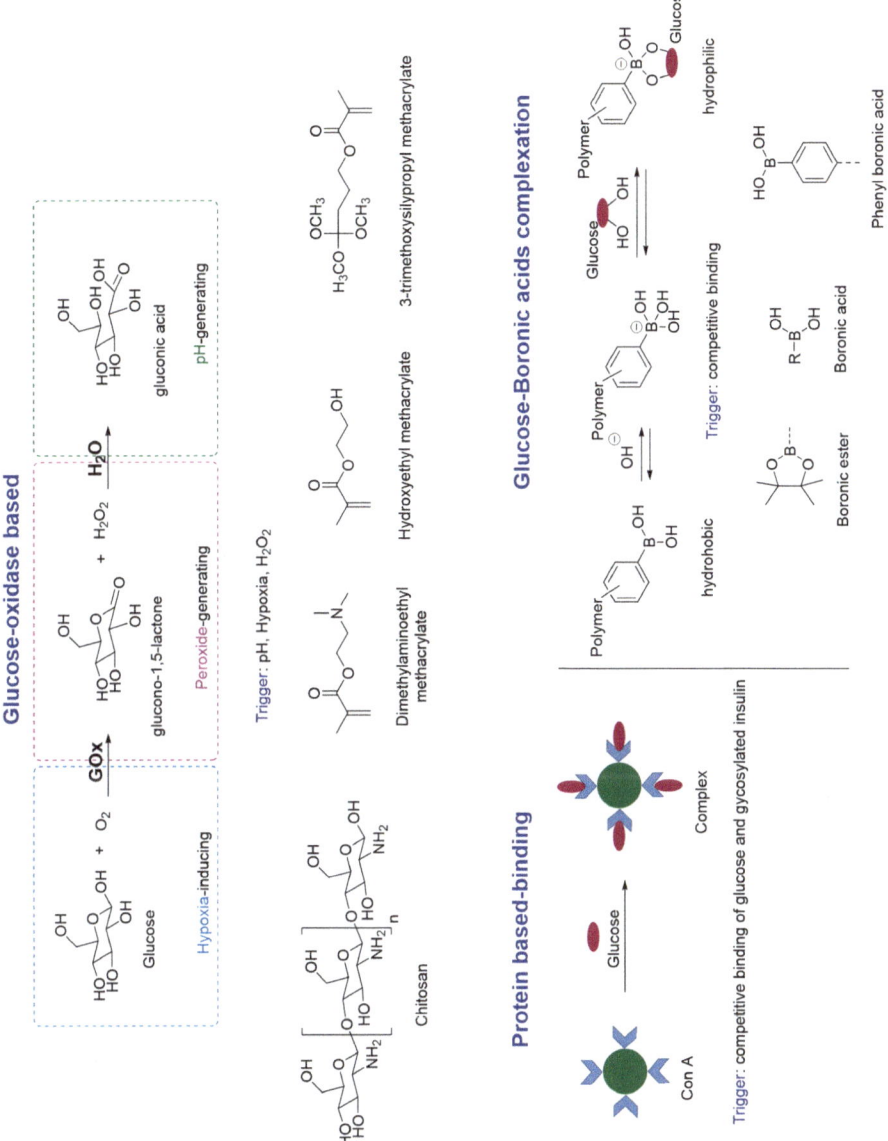

Figure 4.10: Mechanism and chemical structures used in the three main glucose-sensitive polymeric systems.

glucose level grows and is converted into gluconic acid, the pH of the media decreases, and thus the configuration of the polymer is modified, releasing the entrapped insulin [39, 260]. This releasing system mimics the endogenous insulin release.

One of the main advantages of these last two DDS is that they are controlled by a closed-loop feedback system. However, their use is very limited since the proteins are difficult to immobilize, leading to leakages from the polymer and causing a host-immune reaction [198]. Examples of these three types of glucose sensitive polymers used for insulin delivery are summarized in Table 4.9.

Table 4.9: Examples of drug delivery applications for glucose-responsive polymers.

Polymer composition	Drug delivered	System	Discussion	Reference
Chitosan microgel incorporating GOx and insulin	Insulin	Microgel	Glucose oxidase converts glucose in gluconic acid. Chitosan amino groups are then protonated, the matrix is swelled, and insulin is discharged, to control of the blood glucose level	[261]
Copolymer of polyethylene glycol and phenylboronic ester-conjugated polyserine (mPEG-b-P(Ser-PBE))	Insulin	Vesicles	The polymer is H_2O_2-responsive, it is integrated with a transcutaneous microneedle-array patch and contains GOx and insulin. Insulin releases as glucose levels rise, and the kinetics are controlled adjusting the concentration of GOx loaded in the system	[262]
N-(2-(dimethylamino) ethyl)-methacrylamide) and concanavelin	Insulin	Hydrogel	Insulin discharge was reversible according to the glucose concentration. Used for self-controlled insulin release	[263]
Acrylic acid (carbomers (Carbopol 934P and 941P))	Insulin	Gel	Concanavalin was covalently bonded to the carbomers to avoid toxicity. Insulin is delivered in response to the glucose level	[264]

Table 4.9 (continued)

Polymer composition	Drug delivered	System	Discussion	Reference
Poly(styreneboroxole) block copolymer with PEG	Insulin	Micelles and vesicles	The system shows binding to monosaccharides at neutral pH. A polymersome was prepared encapsulating insulin, that can be discharged when sugars are present in the media under physiologically relevant pH conditions	[265]
Poly(3-methacylamido phenylboronic acid) (PMAPBA) and thiolated chitosan	Insulin	Nanoparticles	Disulphide bonds decrease insulin release from the system, and when glucose levels are increased, the insulin delivery increases	[266]

b) Enzyme-responsive polymers

Enzymes play an essential part in the human body controlling biological and metabolic processes, and their capability to bio-recognize, and catalyse physicochemical variations makes them valuable for the development of smart polymer DDS [267]. These systems are ideal when an overexpression of specific enzymes occurs, and when there is a gradient of the concentration of enzymes as a result of a disease, or the combination of both. Enzymes are useful to fix together polymer chains, forming covalently bonded or self-assembled networks; or to break certain bonds causing the rupture of the system. Anyhow, enzymes act on DDS causing the discharge of the drug or modifying its rate [268].

The design of these systems requires an enzyme sensitive component (a substrate for the enzyme). This is achieved by introducing specific enzyme sensitive moieties in the polymer backbone or in the side chains to generate structures that can cleave a particular enzyme, and with the drug physically or chemically entrapped in the system [269]. Then, the DDS needs to be able expose the sensitive groups to the enzyme, so the system has to be designed taking into account the extracellular or intracellular obstacles to achieve enzyme-responsive release.

Hydrolases (proteases, lipases, kinases, phosphatases, etc.) are the most studied enzymes to trigger drug release. They break covalent bonds or modify some specific chemical groups, varying van der Waals forces, steric or π–π, hydrophobic, or electrostatic interactions or hydrogen bonds [270]. A direct release or activation goes through the covalent modification of a polymer with drugs by an enzyme-cleavable linker (Figure 4.11a) and involves a fast release of the therapeutic payload or the activation of the quenched functional agent. For example, proteases can produce drug discharge when it is associated to the polymer by a peptide, glycosidases can act if the transporter is a polysaccharide, lipases when they hydrolyse phospholipid moieties and so on [271, 272]. As an alternative, DDS can be designed to follow an indirect release or activation via physicochemical change (Figure 4.11b). This way, the drug is released by designing the system to act on the polymer main chain through enzyme activity, inducing different physicochemical variations.

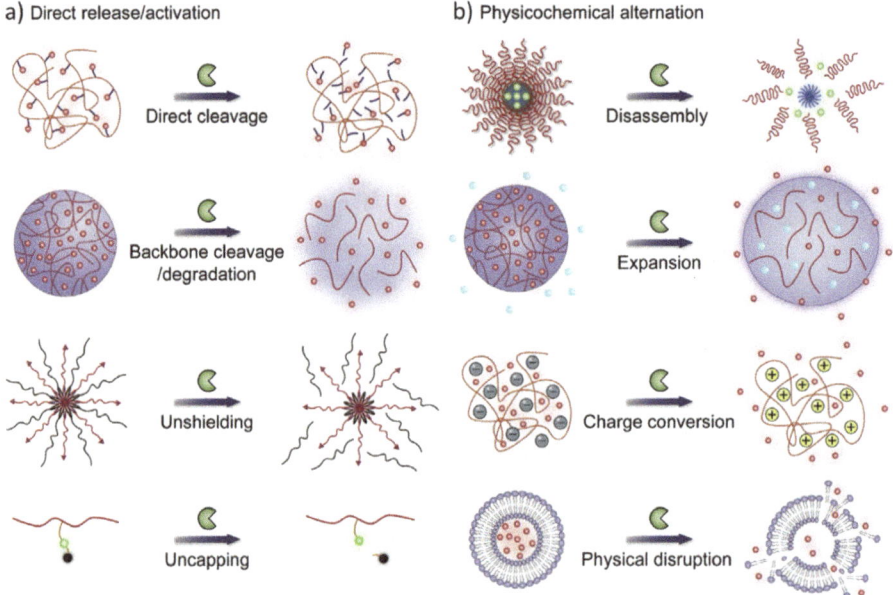

Figure 4.11: Schematic representation of enzyme responsive drug delivery systems following a direct (a) and indirect (b) release/activation. Reproduced with permission of Elsevier [272].

These systems have a number of applications in different diseases. Some examples are discussed in Table 4.10. One of their advantages is their selectivity which reduces their toxicity in healthy tissues. Their main drawback is that they can release the drug before reaching the target location, which can be avoided by the incorporation of other stimuli-responsive properties such as pH.

c) Antigen/antibody-responsive polymers

The treatment of cancer, hypersensitivity and infectious diseases relies on the proper immune response, whose mechanism is guided by antigen/antibody interactions in the body [277]. Antibodies can specifically recognize antigens, and this concept can be exploited to prepare antigen/antibody-sensitive polymers with three different approaches. First, through the combination of the antigen or antibody in the polymer system, through the grafting of antigen/antibody pairs in the polymer carrier as

Table 4.10: Examples of drug delivery applications for enzyme-responsive polymers.

Polymer composition	Drug delivered	System	Discussion	Reference
Chondroitin sulphate (CS) and poly(ethylene glycol) (PEG)	Bone morphogenic protein	Hydrogel	Enables the application-specific tailoring of growth factor delivery and cellular responses. Thrombin hydrolyses peptide amide bonds of fibrinogen	[273]
Poly(methacrylic acid-co-N-vinyl-2-pyrrolidone)	Insulin	Hydrogel	System designed for oral delivery. The polymer shows enzyme-catalysed breakup targeted by trypsin in the small intestine by hydrolysing peptide amide bond at C terminal lysine and arginine	[274]
Azobenzene linked poly(ethylene glycol)-b-polystyrene	Sulfasalazine	Nanoparticles	Micellar disassemble caused by the breakup of azo copolymers after treatment with azoreductases and CoE NADPH. Azoreductases are present in human intestine	[275]
Drug-polymer conjugate, PEG2000-peptide-PTX	Paclitaxel (PTX)	Micelle	Improved antitumour efficiency and targeting by a self-assemble of MMP2-paclitxel micelle	[269]
Poly(ethylene glycol)-b-poly(L-tyrosine) block copolymer	DOX	Nanoparticle	Improved loading capacity and release speed of DOX caused by proteinase K	[276]

cross-linkers or through the physical entrapment of an antibody or antigen in the polymer. The immobilization of biomolecules is discussed in Chapter 3 of this book, so it is not further detailed here. One of the main disadvantages of the first approach, is that the modification goes through the amino group of the lysine moiety, which is a non-specific reaction, and there are several lysine groups, leading to a great modification of the antibody and a concomitant reduction of the sensitivity towards antigens [278, 279]. Despite the high specificity of the antigen/antibody response, it is rarely exploited to prepare DDS due to the high costs, and examples are difficult to find in the literature. These DDSs are mainly used in the form of hydrogels, and some applications are discussed in Table 4.11.

Table 4.11: Examples of drug delivery applications for antigen/antibody-responsive polymers.

Polymer composition	Drug delivered	System	Discussion	Reference
N-isopropylacrylamide (NIPAAm) and *N,N*-methylenebis (acrylamide)	Antibody	Hydrogel	Linking of the antigens to the antibody part bonded to the polymer produced reversible modifications in the swelling	[279]
Polyacrylamide, goat anti-rabbit IgG	Antibody	Hydrogel	When the system is treated with IgG antigen, the cross-linker is lost and swelling of the gel takes place	[280]
Polyacrylamide and vinyl copolymer of goat anti-rabbit IgG	Haemoglobin (model drug)	Hydrogel	Competitive interaction of the free antigen causes modifications in the network volume. Hydrogel shows shape-memory characteristics, and variations in antigen concentration prompts pulsatile permeation of a protein through the system	[281]

4.2.3.8 Multi stimuli-responsive polymers

Most of the existing DDS are based on smart polymers that respond to one single input. However, multi-responsive systems can also be prepared, and they are intended for addressing previous issues. They preserve the primary drug until the objective is reached, enhance drug loading capacity and release and offer further modes of treatment. Multi-responsive systems typically consist of AB or ABC block copolymers, where each of the copolymers is responsive to one particular stimuli. Common combinations include temperature/pH, NIR/pH/redox, enzyme/thermal or enzyme/pH sensitive copolymers [282–284], and some examples are discussed in Table 4.12. The main disadvantage of these systems is that they are difficult to synthesize and thus their applications are limited.

Table 4.12: Examples of drug delivery applications for multi-responsive polymers.

Polymer composition	Stimuli	Drug delivered	System	Discussion	Reference
Poly(ethylene glycol) methyl ether methacrylate, poly(*N*-vinyl-2-pyrrolidone)	NIR, thermal	Bupivacaine hydrochloride	Nanoparticles	Release was higher at 45 °C with and increased with NIR irradiation	[283]

Table 4.12 (continued)

Polymer composition	Stimuli	Drug delivered	System	Discussion	Reference
Poly(ethylene glycol), poly (dopamine)	NIR, pH, redox	Anticancer drugs (DOX and SN38)	Nanoparticles	Improved stability and drug retaining in biological conditions. Multiple stimuli (NIR, light, pH and reactive oxygen)	[284]
Poly(ethylene glycol)	Enzyme, pH	Hydrophobic drugs (i.e. DOX)	Dendritic polymer	Drug release is increased at pH 5.4 when cathepsin B and glutathione are present	[285]
Poly(N-vinyl -2-pyrrolidone)	Enzyme, NIR	DOX	Nanoparticle	Released when hyaluronidase is present, and enhanced with NIR	[286]
Poly (amidoamine) dendrimers and PEG segments	pH, redox	DOX	Polymer conjugates	Solves the contradiction between long circulation time and efficient intracellular drug delivery. Improves antitumour efficiency and present a good safety	[287]

4.3 Tissue engineering

4.3.1 Introduction

Our human body consists of hard tissue such a bones and teeth, and soft tissues, such as muscles, cartilages or skin, containing large amount of water (between 30% and 70%), in a very specific environment that we call "biological media", defined mainly for the pH and water content [288]. These materials in our body combine good mechanical properties (toughness, damping or low sliding friction) with smart behaviour. Indeed, our body presents smart characteristics (e.g. self-healing in the case of wounds or self-wrinkling concerning muscles) [289]. One of the main challenges of researchers involves the use of materials with similar properties to human tissues [290]. In this sense, the search of these type of materials is large, covering from biomaterials extracted from the body itself [291] to new rubber-based nanocomposite materials [292], with the objective of "developing biological substitutes that restore, maintain, or improve tissue function or a whole organ", according to the classic work of Langer and Vacanti [293].

Tissue engineering involves four components, as depicted in Figure 4.12. Firstly, progenitor or steam cells; then, a scaffold material that can be natural or synthetic that serves as architecture for adhesion, transplantation and cell function; a bioreactor to provide an active media for cell proliferation; and finally, signalling molecules to bind receptors.

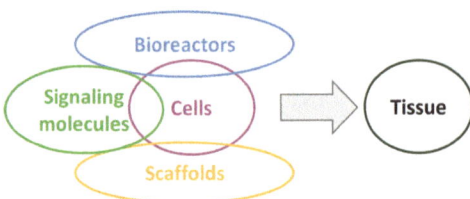

Figure 4.12: Components of tissue engineering.

Simple polymers can provide the mechanical and physical support needed for the development of tissue. However, they lack biomimetic properties and interactions with human tissue, which is still limiting for functional tissue formation. For this reason, efforts are being made to develop advanced smart materials able to respond to stimuli and thus interacting with human stem cells together with providing structural integrity. In this sense, smart functional polymers, both synthetic and natural, are able to control physical, chemical and biological properties. In addition, the versatility of these materials (they can be synthesized from different monomers, including bioactive molecules) results in materials with different compositions, molecular weight, shapes and properties, and thus they can serve for tissue engineering [293]. Accordingly, smart cross-linked hydrophilic polymers forming a 3D network, smart hydrogels, have emerged as the ideal "biological substitutes" to replace human tissues. The polymer provides mechanical and porous network structural support, with surface chemistry for cell attachment and growth, cell–cell communication (through the porous structure) and cell proliferation and differentiation for tissue development. In this sense, the smart function of the polymers is directed to maintain and control cellular behaviour for functional tissue regeneration. As mentioned previously in this book in **Section 4.2,** smart-responsive polymers vary their properties as a response to one or more stimuli, including pH, temperature, redox potential, light electric or magnetic field, biomolecules and so on, which influences cell behaviour and functionality, and this way, smart polymers for tissue engineering can respond to body variations as a functional tissue would do.

Following these ideas, functional tissue can be developed both *in vitro* and *in vivo* in a surgical procedure, and they are crucial for injury repair.

There are different approaches to tissue engineering, as described in Figure 4.13. Normally, tissue-specific cells are isolated from the patient and harvested *in vitro*. In this approach, these cells are expanded and seeded into hydrogels scaffolds that mimic the extracellular matrices (ECMs) of the damaged tissues. The functions of the hydrogels include the delivery of the seeded cells to the target site of the patient's body and to promote cell–biomaterial interactions and adhesion and enhance the transport of gases and nutrients to ensure cell survival, among other functions. In a second step, hydrogels scaffolds are transplanted into the patient by two main techniques: Through surgery, with all the negative implications associated to the patient, and a second one, which has been employed successfully in the

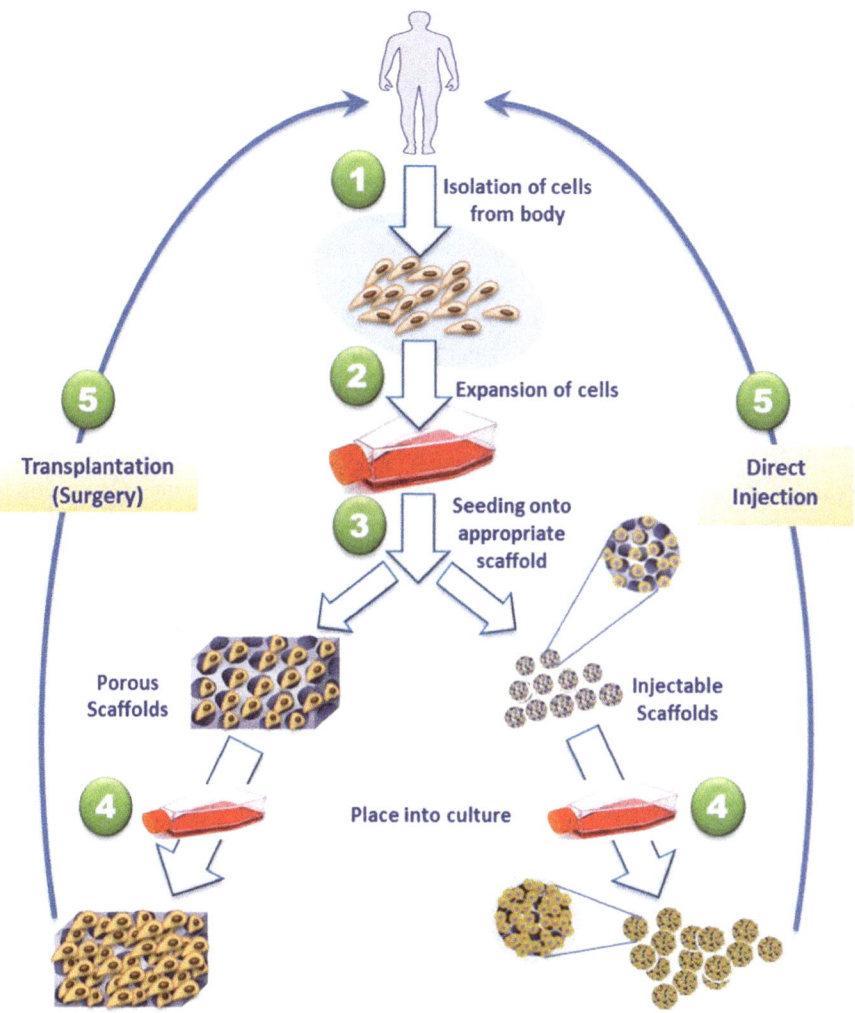

Figure 4.13: Tissue engineering approaches. Tissue-specific cells are isolated from a small biopsy from the patient (1), expanded *in vitro* (2), seeded into a well-designed scaffold (3) and transplanted into the patient either through injection (4, right), or via implantation at the desired site using surgery (4, left). Reproduced with permission of Bloomsbury Qatar Foundation Journals [296].

last years, using injectable hydrogels [294]. The importance of injectable hydrogels has grown exponentially in the last years, because they present minimal invasive administration and they can be mixed homogeneously with cells and therapeutic agents [295, 296]. Some of the latest advances associated to the use of smart injectable hydrogels in tissue regeneration are described in this chapter.

Focusing our attention on these materials and their importance in tissue engineering, the first part of the section will define what smart hydrogels are and their essential characteristics and the materials used to prepare them, then, the application of stimuli-responsive hydrogels and the most recent applications of these materials in tissue engineering will be covered.

Hydrogels are one type of polymeric materials, consisting of 3D networks of cross-linked and hydrophilic polymers, which have the ability to hold very large amounts of water, showing an outstanding swelling behaviour [297, 298]. There are different characteristics that define the behaviour of a hydrogel: swelling ratio, defined as the percentage of water uptake, the gelation time, the porous structure, in terms of pore size and morphology (closed or open) and also the weight ratio of polymer, which can be extremely low (from 0.1% to 10%) [299]. The ability to control the surface morphology, shape, size and porosity of the hydrogel has enabled the overcome of a number of challenges in tissue engineering, such as vascularization. In general, polymeric hydrogels for biomedical applications are usually referred to as "scaffolds" [296]. The presence of an open porous structure is essential [300] because it enables the transfer of nutrients and cellular waste. Finally, hydrogels can be obtained from biocompatible and biodegradable polymers [301], which allow the control of the migration and growth of cells during regeneration [302].

These materials can be produced by different polymerization techniques, aimed to obtain a cross-linked hydrophilic network to produce an elastic structure. The preferred procedure is the conventional free radical polymerization when polymerizable functional groups are present. Some other techniques include the cross-linking of a polymeric precursor by irradiation, chemical cross-linking or even physical cross-linking through polyelectrolyte complexation, hydrogen bonding or hydrophobic association. In terms of polymerization techniques, any classical procedure can be employed to form hydrogels, such as bulk, solution and suspension polymerization [303].

Hydrogels can be classified into natural or synthetic according to their origin; as durable or biodegradable; and they can also be classified according to the response to stimulus (smart- or stimuli-responsive hydrogels). Smart hydrogels can go under reversible changes in swelling behaviour, network structure or mechanical properties in response to various external factors, including temperature, pH, light or electric and magnetic fields [304–306]. The volume phase transitions can be easily triggered, then obtaining smart hydrogels in which their cross-linking and swelling behaviour can be tuned, making them attractive to fulfil the requirements of tissue regeneration [307].

4.3.2 Materials for smart hydrogels

As described in this chapter's introduction, hydrogels provide a series of different functions in tissue engineering to mimic ECMs. These functions include their use as structural integrity and bulk for cellular organization, act as tissue barriers and provide adhesion, serve as drug carriers, encapsulate cells and provide bioactive moieties for natural healing:

- Hydrogels can serve as carriers for cell transplantation as they provide unique immunoisolation and at the same time they allow the diffusion of nutrients, oxygen and metabolic species.
- Hydrogels serve as scaffolds. Their physical and mechanical characteristics can be tuned to mimic natural tissues, where cells can be suspended or adhered.
- Hydrogels are barriers against restenosis, showing improved healing response after surgery procedures. Hydrogels act as thin films barriers to avoid contact of vascular walls with platelets, plasma and coagulation factors.
- They can be used as localized drug depots due to their hydrophilicity, biocompatibility and stimuli-responsive ability. This ability can be used simultaneously with the barrier function and the inhibition of post-operative adhesion formation.

There are no ideal hydrogels that can be employed in all tissue engineering applications. It is necessary to select the properties as a function of the cell type, tissue or pathology. In this sense, it is a very different application in which the hydrogel is used as a permanent scaffold for the growth or replacement of a bone from a temporary hydrogel that supplies a reservoir of cells in nerve regeneration. Then, the first step in tissue regeneration engineering, before selecting the ideal polymeric hydrogel, is to take into account several key considerations [299]:

- Will the hydrogel be used *in vivo* or employed to grow tissue *in vitro*?

It is important to consider that polymer scaffolds can be tolerated in the laboratory by isolated cells using controlled conditions, but when they are implemented into the body, environmental conditions change (and also, they are exposed to many cell types which could affect drastically their performance [308]). Also, a hydrogel implanted *in vivo* into a patient is mechanically stressed from a very different manner than a hydrogel used *in vitro*, which defines the type of polymer that must be used [309].

- Must the hydrogel behave as a space-filling material or to fill a defined 3D architecture?

This design point is related to injectable hydrogels that are essentially used in *in vivo* applications, and more specifically, to their gelation behaviour. For example, in free filling space applications, it is usual to use preformed smart hydrogel in which different stimuli (pH or temperature) triggers a fast and *in situ* gelation upon injection [310]. In 3D patterns, the gelation must be even faster, and new applications in the design of 3D printing and stereolithography using cell-containing "bio-inks" are being explored [311].

– Which is the useful lifetime of a hydrogel?

This parameter is referred to the type of application in which the hydrogel is used. Some hydrogels are applied as permanent implants, where others (more commonly) are gradually removed from the site after their healing action. Another interesting possibility is the recent use of hydrogels with degradable cross-links, allowing the development of bio-resorbable polymers. The hydrogel is de-cross-linked, obtaining small monomers that can be easily eliminated from the human body [312]. A well-known example of biodegradable polymer is the hyaluronic acid (HA), which is rapidly digested by endogenous enzymes, but leading to very low tissue retention times [313].

– Should we employ an inert hydrogel or an active hydrogel in which proteins can adhere?

In most of tissue engineering applications, the hydrogel provides an adhesive scaffold for cell binding. Then, it is necessary to mimic the recognition motifs present in the ECM to bind proteins and carbohydrates. Binding synthetic peptide sequences usually achieve this to enhance cell binding, for example in PEG-based hydrogels [314, 315]. However, although it is important to remark that synthetic hydrogels present a bioinert nature, this behaviour can also be used advantageously to control and limit cell attachment and growth, because natural polymers can induce undesired signalling.

4.3.2.1 Natural polymers

The importance of hydrogels based on natural polymers lies in their macromolecular similarity to the physical and chemical conditions of the ECM. One of the main advantages of using natural polymers is that the degradation products are non-toxic and can integrate easily into the human body. On the other hand, these polymers present rapid degradation and poor mechanical properties [316, 317].

Natural polymers employed as hydrogels are essentially divided in two groups: polysaccharides and proteins. Polysaccharides present a resemblance to the biological ECM through the presence of glycosaminoglycans, which are long linear polysaccharides consisting of repeating disaccharide units [318]. Therefore, we can remark different polysaccharide-based scaffolds based on natural polymers such as:

– Chitosan, a natural polysaccharide used extensively due to its low toxicity, biodegradability and biocompatibility. In terms of their performance, chitosan can be easily degraded by lysozyme, a protein naturally present in the human cartilage [319].

– HA, or sodium hyaluronate, is a linear non-sulphonated polysaccharide, which is also present in the ECM. HA can affect cell proliferation, inflammation and wound reparation. This natural polymer can also support cartilage formation through chondrogenic differentiation of mesenchymal stem cells [320]. HA has attracted great attention in the recent years in cosmetic applications as rejuvenating skin agent [321], but it also has several disadvantages related to the limitation as free radicals, being necessary to introduce different functional groups such as –OH, –COOH or N-acetyl groups to increase their durability and mechanical properties. Another key factor is

the molecular weight of HA, which defines the hydrogel´s application, from blocking endothelial cell migration to anti-inflammatory properties [322].
- Alginate is a natural polysaccharide combining mannuronic acid and guluranic acid blocks. This chemical structure allows using alginate in hydrogels combined with Ca(II) and Ba(II), which are inert immunologically and can be applied in the regeneration of mammalian cells [323]. However, one of the main drawbacks of alginate is related to the degradation process, which depends essentially on the diffusion of cations, and cannot be controlled properly [324].

The second group of natural hydrogels is based in fibrous proteins, which have attracted many attentions in the last years because they greatly resemble the ECM, in terms of structure and mechanical properties, then allowing cell migration and growth easily. In addition, this natural hydrogel type is highly hydrophilic and can be easily cross-linked chemically or physically [325]. Among other protein-based hydrogels (such as elastin or keratin), the most important biopolymers are collagen, gelatin and silk:
- Collagen is the major and most important fibrous protein in connective tissue, and presents many advantages such as biodegradability, biocompatibility, easy availability and great versatility, although again, its poor mechanical properties can limit their application fields [326]. There are, essentially, five different collagen types depending on the tissue in which they are present. Type I (skin, tendon, vasculature or organs), type II (cartilage), type III (reticular fibres), type IV (epithelium-secreted layer of the basement membrane) and type V (cell surfaces, hair and placenta) [326].
- Gelatin is the denatured form of collagen, obtained from broken triple-helix of collagen. These hydrogels are thermo-responsive, with an UCST between 25 °C and 35 °C, but they need to be chemically modified to enhance their poor thermal and mechanical stability and to obtain materials with long durability [327].
- Silk fibres are derived from natural silkworms, and they are formed by core filaments of silk fibroin protein and a rubbery coating protein named sericin [328]. In this case, these polymers also present good biocompatibility and processability, combined with excellent mechanical properties, making them a very promising material for building scaffold structures. Also, it can be easily degraded by ubiquitous proteases with different ratios depending on the formulation of the silk protein [329].

As said before, other protein-based hydrogels can be obtained starting from fibrin (a natural component of blood clots, which make them very useful in endothelialized tissue engineering [330], and keratin, the main fibrous protein in hair or wool, with great stability and special physicochemical properties due to the intermolecular bonding of disulphide cysteine and amino acids [331]. Figure 4.14 lists some of the chemical structures of different natural polymers employed as hydrogels.

4.3.2.2 Synthetic polymers

One of the main drawbacks of natural polymers is their rapid degradation and low mechanical strength, limiting their applicability in tissue engineering. This issue

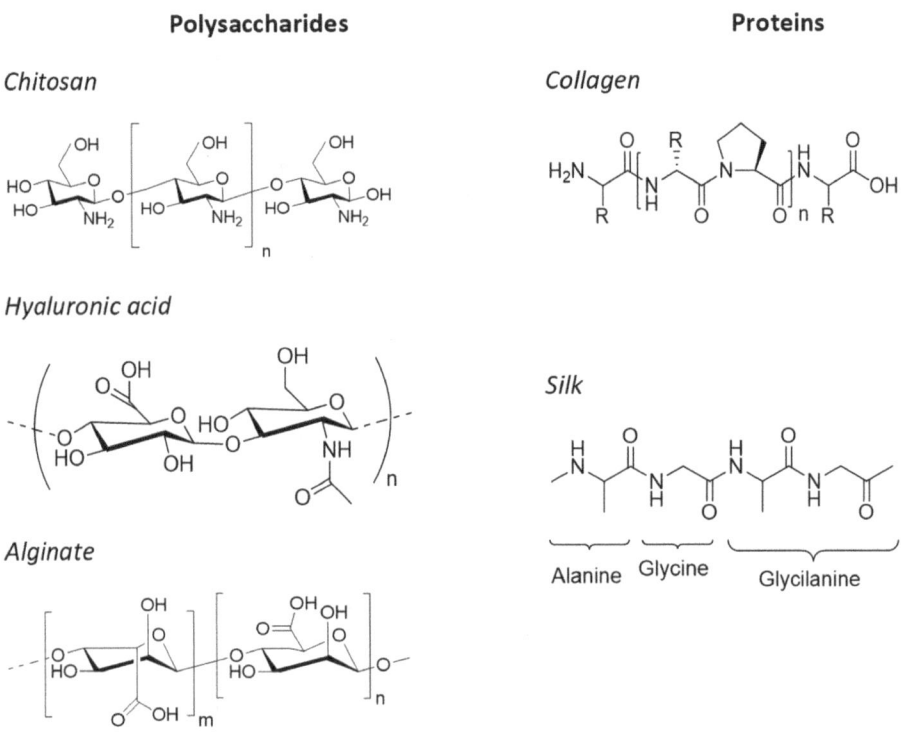

Figure 4.14: Examples of natural polymers employed as hydrogels in tissue engineering.

has motivated the research around synthetic hydrogels, which can be tuned physicochemically to overcome these difficulties. On the other side, the biological activity of synthetic polymers is much lower because they tend to increase their local pH by acidic degradation products and present poor degradation. However, these materials present the possibility of controlling their macrostructure, porosity and mechanical properties to widen the applicability range of natural hydrogels [332, 333]. Some of the most important synthetic polymers employed as smart hydrogels in tissue engineering are shown in Figure 4.15, and depicted below:

- PNIPAAm is a thermo-responsive polymer with a LCST temperature around 33 °C, close to the temperature of the human body. This hydrogel can trigger its swelling behaviour together with its mechanical stiffness by varying the ambient temperature above or below this LCST. The formation of a hydrogel is based on the

copolymerization of PNIPAAm with hydrophilic and hydrophobic monomers to obtain the desired properties in biomedical applications [334, 335].
- Poloxamers are triblock copolymers of a hydrophobic core of propylene oxide and two hydrophilic units of ethylene oxide. The importance of these copolymers is that they are commercial (trademarks Pluronic®, Synperonic® or Tetronic®) [336].
- Poly(vinyl alcohol), or PVA, is a synthetic hydrophilic polymer that can be easily cross-linked [337]. PVA is usually mixed with collagen to enhance cross-linking efficiency, obtaining smart hydrogels with good biocompatibility, good degradation behaviour and low toxicity [338].
- PEG is a hydrophilic polymer with a backbone likely to form hydrogen bonds with water [339]. Also, PEG is biocompatible with proteins and enzymes, and it can be used in many biomedical applications through the coupling of PEG macromeres with peptides and similar ECM molecules [340].
- Polycaprolactone or PCL is a polyester-based polymer with many advantages in tissue engineering applications [341]. First, it is easily synthesizable and processable, it shows excellent compatibility and it can be mixed and copolymerized with natural polymers (collagen or HA) to obtain hybrid materials with good mechanical properties for biomedical applications [342].
- Polylactic acid or PLA is a biocompatible synthetic polymer that has been used (employing porous structures) in the regeneration of different tissues [343]. PLA is also easily biodegradable, and it can easily incorporate cell cultures for *in vivo* applications. PLA-based hydrogels may also be tailored as stimulus-responsive (pH, photo or redox) materials by copolymerization with PEG or other synthetic polymers [344].

4.3.3 Stimuli-responsive hydrogels in tissue regeneration

Stimuli-responsive polymers were previously described in this book regarding their use for drug delivery applications, and thus, only specific applications of these materials, especially hydrogels, on tissue engineering are described in this chapter, including mechanical responsive materials [345].

Temperature-responsive polymers were described extensively in **Section 4.2** in this book. Temperature-dependent change is related to the hydrophilic/hydrophobic balance in the network [345]. When temperature is increased/diminished above/below the CST of the polymer, disruption of hydrogen bonding with water allows intra- and intermolecular bonding and hydrophobic interactions to dominate, then aggregates the polymer chains and lowers the swelling ability of the hydrogel [11]. The importance of these smart hydrogels in tissue engineering lies in their ability to reach the gelation point easily without needing any other chemical or physical phenomena, only induced by changing the environment temperature. Then, hydrogels with a LCST lower than human body temperature can be used as injectable systems [346].

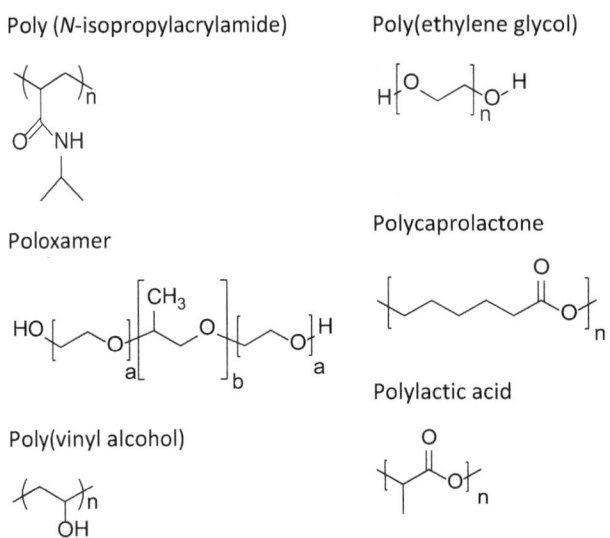

Figure 4.15: Some synthetic polymers used as hydrogels in tissue engineering.

Some examples of smart hydrogels for tissue engineering are based on *N*-isopropylacrylamide (NIPAAm), which present a LCST around 32 °C, and it is frequently used in the fabrication of thermo-responsive surfaces, in which surface roughness is easily regulated to tune the attachment (above the LCST) or release (below the LCST) of diverse cells in cell culture applications [347]. Also, NIPAAm can be combined to methylcellulose to obtain thermo-reversible hydrogels employed in cartilage regeneration [348].

Another common type of stimuli-responsive polymers used in tissue engineering are pH-responsive hydrogels, that present acidic or basic groups which exchange protons depending on the pH, through a mechanism of disassociation and association with hydrogen ions when pH is varied [349]. This type of hydrogel is employed mainly in drug delivery applications, as described previously in this book (**Section 4.2**), in which the pH profile of different pathologies, such as infections or cancers, is very different from the pH of healthy tissue [350]. PAAc is the classical example of weak polyacid that can be protonated and deprotonated by varying the pH, and it is used in the preparation of block copolymers to obtain pH-responsive hydrogels [351]. These hydrogels can also be employed in tissue regeneration, examples can be found in the literature in which chondroitin sulphate PEG adhesive pH-responsive hydrogels were synthetized to regenerate cartilage tissues [352], and also in which pH-responsive injectable hydrogels based on chitosan and hydroxyapatite were synthetized, proving their ability to promote angiogenesis, a key process in bone regeneration [353].

Also, the use of light to tune hydrogel properties has also become very interesting in the last years. It is necessary to incorporate photosensitive functional groups to the hydrogel, such as photochromic chromophores (examples can be found in **Section 4.2** of this book, to obtain phototriggered hydrogels [354]). The use of light stimulation presents several advantages with respect to other external stimuli in tissue engineering: the capability of contact-free remote manipulation of biomaterial properties, the possibility of fine adjusting of wavelength and intensity and, more importantly, water is almost transparent for light in the photochemically relevant range (NIR-UV), improving the efficiency and functionality of hydrogels as scaffolds in biological media [355]. These hydrogels can also be used for tissue reparation, especially cartilage regeneration, following a simple procedure. The regeneration cells are encapsulated in the non-polymerized hydrogel, which is subcutaneously injected into the damaged tissue, and then transdermal photopolymerization is carried out to obtain a cross-linked material that acts on the targeted cartilage by releasing the healing cells. Networks based on PEG and PVA with encapsulated chondrocytes can be prepared to repair neocartilages [356], and also biodegradable materials, such as gelatin or HA, can be used to form photocurable hydrogels based on methacrylamide copolymers to repair cartilage tissue [357].

Electrosensitive hydrogels are usually composed of electrolytes, and their behaviour is similar to pH-responsive hydrogels (see also **Section 4.2** for further description of these type of polymers). In this case, the properties variations (swelling, shrinking or mechanical) are tuned by the presence of an electric field, and more specifically, they depend on the orientation of the hydrogel with respect to the electric field, and also on the intensity and duration of the stimulus [358]. This type of hydrogels has been employed in different research works related to tissue regeneration. For example, poly(acrylic acid)/fibrin electroactive hydrogels have been successfully tested to promote collagen production through three-dimensional culturing, conditioning smooth muscle cells [359] and, recently, neural tissue engineering using electroactive hydrogels based on chitosan–aniline oligomers/agarose has been developed with very promising results [360].

Mechanically responsive hydrogels react when external mechanical stresses are applied. In this sense, mechanical forces can stimulate or trigger smart hydrogels to control growth factors in biomedical applications [361]. One important key point to consider is that since musculoskeletal tissues are constantly under mechanical stressing, it is essential to evaluate the influence of these stimuli in the cellular behaviours and biological parts [362]. Then, mechanical forces can play a key role in regulating cartilage development, specifically in the chondrogenic process [363]. Some research works have been developed using mechanical-responsive hydrogels in tissue regeneration, although this type of materials are not being extensively used. Hydrogels based on 2-acryloylamido-2-methylpropane sulphonic acid can be used for stem cell culture, showing that the matrices support survival, proliferation and differentiation of stem cells [364]. Also, hydrogels based on nanofibrils of PVA have been tested as artificial muscle-like fatigue-resistant materials [365] to work as possible substitutes of damaged natural tissues.

Hydrogels with the molecular recognition capabilities have found great applicability in different biomedical areas, such as insulin delivery [366], blood coagulation [367] and protein immobilization [306]. In terms of tissue engineering, there are some examples of bioresponsive hydrogels with interesting regeneration properties, mostly enzyme-responsive hydrogels. A representative example is the development of an enzyme-sensitive hydrogel based on PEG cross-linked with a peptide derived from aggrecanase, which showed the ability of cartilage repair [368]. Concerning the other applications, they are described in more detail in **Sections 4.2** (drug delivery) and **3.6** (protein immobilization).

Table 4.13 resumes the characteristics of responsive hydrogels employed in tissue engineering. Although we have treated each stimulus separately, we must remark that different efforts have been developed to obtain multi-responsive smart hydrogels. Material properties can be triggered by different external stimuli, obtaining versatile materials that can be used in different conditions and environments [369].

Table 4.13: Examples of stimuli-responsive hydrogels used in tissue engineering and governing mechanisms of behaviour (adapted from [302]).

Stimuli	Mechanism	Materials
Temperature	Changes in swelling behaviour	NIPAAm, pluronic, PEG copolymers
pH	Association/dissociation of pendant acidic/ basic groups with hydrogen ions	PAAc, chitosan, poly(N-vinylcaprolactam)
Light	Incorporation of photosensitive functional groups into the hydrogel network	Azenobenzenes, poly(cinnamic acid), triphenylmethane
Electric field	Generation of electrical polarization in polyelectrolytes	Chitosan/PAN, hyaluronic acid/PVA, alginate/PMMA
Mechanical stress	Induced conformational changes	Peptides, acrylate-based block copolymers
Biomolecules	Dissolution/precipitation dynamics	Glucose-sensitive, DNA-responsive, enzyme-sensitive hydrogels

4.3.4 Recent developments and applications

We can classify tissue engineering applications into two main categories *in vitro* and *in vivo* tissue engineering. The first category corresponding to *in vitro* tissue engineering is carried out outside the patient's body, to develop a functional tissue construct prior to implementation [370]. Following this approach, cardiac tissues can be developed in culture dishes using engineering scaffolds to apply electrical stimulation in a second step [371]. Another example of successful *in vitro* tissue engineering is developing an artificial bladder, starting from biopsies from patients (autologous tissue engineering) that were grown in culture and seeded into a collagen–poly(ethylene) glycol hydrogel. These structures were later implanted into patients to restore their bladder function [372]. A more complex reversible-responsive fluorochromic hydrogel, suitable for stem cell development, is based on a lanthanide-mannose complex [373]. In this case, in real-time, fluorescence labels were used to develop gelatin-based 3D cell culture. Another *in vitro* research line uses stem cells to derive specific cell types, which must be clearly differentiated before implementation. Using this method, a human embryonic stem cell-derived pancreatic endoderm can be generated [374]. Stem-cell culture can also be used to develop hematopoietic cells using PVA-based hydrogels in culture fluids, carrying out *in vivo* tests using live mice [375].

The second category, called *in situ* or *in vivo* tissue engineering, is based on the recruitment of endogenous stem cells to the site of injury, using the native regenerative potential of the body itself. In this approach, it is necessary to enhance the healing process through different biocompatible hydrogels, which play a key role in these procedures. This approach requires considering the extracellular factors, due to the direct interaction of the ECM with the hydrogel, which can lead to the immune response of the body or even its rejection [376]. Then, the use of the European Medicines Agency (EMA), the Food and Drug Administration (FDA) or other governmental agencies approved materials is compulsory for clinical use. In this sense, natural polymers do not present any issue (such as protein-based materials as silk or elastin), and concerning synthetic polymers, PLA, PLGA or PLC can be used *in vivo* applications. Different *in vivo* applications can be found in the literature, from the engineering of human skeletal muscles [377] and the reparation of the physics, a cartilaginous tissue in children's long bones that is responsible for bone elongation, using alginate – chitosan-based hydrogels [378].

Among all the types of hydrogels used in tissue regeneration processes, injectable hydrogels have emerged as the ideal candidates to lead this research line, due to several advantages such as possibility to repair a specific damaged area, they can be used in minimally surgical invasive procedures, or they can be modified with nanocharges to improve their performance [379]. First, the building blocks of the hydrogel preparation are collected, including, if necessary, biological cells from the human body. Then, the hydrogel is formed *in situ* through cross-linking (physical or chemical) to capture the biological compounds which will act as repairing agents (biological compounds, drugs, growth factors, etc.). Finally, the hydrogel is injected at the site of interest. The importance of polymeric injectable hydrogels has grown exponentially in the last years [380], and for this reason, we will focus our description of the latest applications in this type of materials.

Two main factors define the targeting region of smart polymeric injectable hydrogels in tissue engineering. First, the mechanical behaviour, related to tissues involving bones and muscles, and second, the electrical properties, which are essential in cardiac or nerve tissues [381]. Then, it is usual to separate the applications of these hydrogels in musculoskeletal, cardiac and neural tissue engineering.

Musculoskeletal tissue engineering is related to bone and muscle (cartilage) reparation. Focusing our attention on bone regeneration, we must remark that bones are the second most transplanted tissue in the world [382], and for this reason, the use of hydrogels in bone reparation has gained a lot of importance through the use of stem cells and scaffolds, able to create an environment in which the formation of new bone tissue is possible [383]. The most important mechanism of hydrogels in tissue regeneration is based on electrostatic forces, where different ions can interact with oppositely charged ligands of the polymeric chains. This mechanism is especially interesting when the smart hydrogel responds to the presence of Ca(II), naturally present in the bone [384]. Recent studies have shown the possibility of using smart hydrogels of HA modified with biophosphonate, which can bind reversibly to Ca(II) ions, obtaining an injectable hydrogel with excellent healing properties, but low mechanical properties and poor stability at physiological pH [385]. Another recent and innovative feature applied to bone regeneration is a smart hydrogel based on protein DNA, which can cross-link through DNA hybridization. In this case, gelation occurs through the addition of a complimentary multi-arm DNA cross-linker. This mechanism allows the design of customized hydrogels that could remedy the bone resorption caused by osteoporosis disease [386].

Muscle (or cartilage) reparation has also become an essential research field in tissue engineering due to the increasing number of injuries occuring in the general population caused by trauma or sport-related problems. The main problems associated to cartilage reparation are the absence of vascularization and the low activity of the ECM [387]. Moreover, the mechanical properties of the hydrogel are especially crucial in this application. For these reasons, using nanomaterials as reinforcements of smart hydrogels has become an interesting alternative to classical hydrogels, showing an attractive approach to tailoring the hydrogels' mechanical properties [388]. Additionally, using electrically conductive charges, such as carbon nanotubes, enhance the electrical conductivity of hydrogels (needed in muscle regeneration applications), also increasing their responsive behaviour. To overcome this issue, there is a new investigation line concerning the use of hydrogels based on conductive polymers, combining the redox switching capabilities of conductive polymers with the fast ion mobility of hydrogels [243]. Then, different efforts have been carried out to obtain smart hydrogels based on PPy, PAni and PEDOT. However, the main disadvantage of these polymers is their biocompatibility, which limits their applicability only to specific tissue types, such as bone (in the case of PPy) [389], neural tissue (for PAni-based hydrogels) [390] or epithelial cells (in PEDOT hydrogels) [391].

Skin is the largest organ of the human body, and it is our first defence line against all type of pathogens and external aggressions. However, it has also a very delicate nature which is easily damaged. In addition, skin transplantation is a very complicated procedure that is usually avoided. Then, tissue reparation using regenerative wound dressings have gained importance in the last years [392]. The most important factor to consider in skin-related applications is designing suitable hydrogels because of the different growth factors involving their two main layers (epidermis and dermis). A schematic representation of the skin layers is presented in Figure 4.16, in which three different growth factors are presented. Epidermal growth factor and keratinocyte growth factor control the epidermis regeneration process, and finally platelet-derived growth factor activates the immune cells and induces fibroblast differentiation in dermis layer [393].

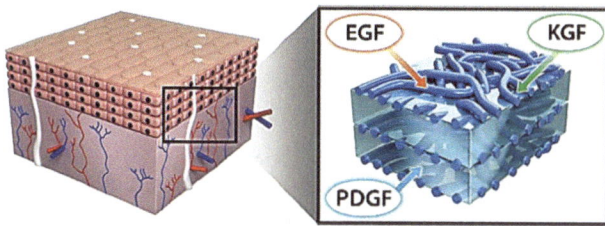

Figure 4.16: Schematic representation of the skin layers (epidermis in red and dermis in blue), and the different skin growth factors involved. Reproduced with permission of Springer [393].

In this field, hydrogels play a key role as healing wound dressings, because they can maintain the environmental moisture around the wound site, can easily absorb the wound exudates and also guarantee a good oxygen permeability to enhance wound tissue regeneration. Most of these hydrogels are based on membranes or films from smart polymers directly placed in the wound to regenerate the damaged tissue (for example, hydrogel membranes based on PVA have been analysed in wound repair or even skin substitutes [394]. Another research work presented, for example, the development of an adhesive and conducting wound dressing based on chitosan-g-polyaniline functionalized with PEG. Cross-linking could be easily triggered thermally at 37 °C, also obtaining an injectable hydrogel. This material has been tested comparatively against commercial products (Tegaderm®) in wound dressing of mice, showing a drastic improvement of the healing properties. The authors believe that the unique combination of electrical conductivity of the PAni and free radical scavenging of the chitosan allowed to obtain a durable material (up to 15 days) with excellent wound healing ability [395]. The encapsulation ability of the hydrogels can also be employed to release natural anti-inflammatory substances, such as curcumin, which can be released *in situ* using thermo-responsive hydrogels based on PEG, acting directly in the wound site [396]. The incorporation of drug nanoparticles into the hydrogel structure has also been analysed, incorporating polydopamine nanoparticles into a PNIPAAm, obtaining a hydrogel responsive to IR radiation, good mechanical properties and great healing ability, tested in an *in vivo* skin-defect model in mice, completely repaired after 15 days [397].

One of the most important death causes is myocardial infarction, and for this reason, the treatment of cardiac related injuries is a worldwide health problem. Among many other solutions, injectable hydrogels have attracted a lot of interest to restore the damaged myocardium membrane because they are practically non-invasive techniques and provide a high regeneration capacity [398]. However, it is important to remark that heart issues are located in a human body region in which pulsatile movements cause stress that could damage the hydrogel itself or the delivered therapeutic materials. Some of the most recent research works include, for example, the fabrication of injectable pH-responsive hydrogels based on PEG with ureido-pyrimidinone moieties. The hydrogel is basically fluid at basic pH, facilitating the injection process, but it is reversibly cross-linked at neutral pH (reaching the heart). The hydrogel was able to reduce the size of a myocardial wound in a pig heart, then reducing drastically the risk of an infarction [399]. Other research work investigates the obtention of a conductive hydrogel based on dibenzaldehyde-terminated PEG copolymer with aniline tetramer incorporated. This material was able to encapsulate C2C12 myoblasts and cardiac myocytes, delivering these cells rapidly into the myocardial membrane, and showing a very fast degradation profile, being completely reabsorbed over 45 days without any inflammatory or self-immune reaction [400].

The last type of target tissue that we will briefly describe is related to the central nervous system (CNS) (neural tissue). In this point, this tissue engineering approach is closer to the cell therapy procedures that we will describe in **Section 4.4** of the book, but we will introduce some guidelines in this chapter. In CNS reparation, the use of hydrogel-based scaffolds, instead of injectable hydrogels, is ideal because they reduce drastically the risk of cell damage due to the injection process, and also, they present a minimally invasive cell-engraftment for the patient (engraftment process occurs when the blood-forming cells received on transplant start to grow and make healthy blood cells) [401]. Also, in this specific research field, special self-healing hydrogels are used, because they enable *in situ* gelation without the need of any environmental trigger [402]. However, these hydrogels present weak mechanical properties (shear moduli lower than 50 Pa) compared to the mechanical strength of CNS tissue (up to 10 times larger). Then, recent developments include the use of reinforcements, such as amyloid nanofibrils which enhance the mechanical strength or hydrogels [403], or the use of biobased self-healing hydrogels, such polysaccharide-based materials or chitosan-based hydrogels, with stiffness up to 1 KPa, close to the value observed in neural cells [404, 405]. To conclude, some efforts are being carried out introducing graphene into the hydrogel matrix, to obtain materials with electrical properties and self-healing capacity, able to, for example, support the growth of neural like P12-cells [406].

4.4 Precision medicine and cell therapy

4.4.1 Introduction

This last part of the chapter is dedicated to the role of smart polymers in new biomedical applications related to precision medicine (or nanomedicine) and cell therapy, mostly devoted to cancer treatments and diagnosis of important diseases such as Alzheimer or Parkinson. The most relevant advances found in the literature in this sense will be discussed. Following this scheme, this chapter is divided in two main parts:

- *Precision medicine:* Precision medicine, also called personalized medicine, relies on personal clinical, genetic and environmental conditions. Smart biocompatible and adjustable polymeric materials allow the development of precision medicine as smart polymeric materials can be designed to each own specific individual. Polymeric nanomedicine, is referred to the use of nanosized nanoparticles or conjugates (1–1,000 nm diameter), used for biomedical applications (diagnosis and treatment) of diseases, both *in vitro* and *in vivo* [407]. The main concepts and latest applications of non-invasive surgery, the use of microfluidic devices in biomedical applications and the recent developments in the use of 3D bioprinting processes using smart polymeric materials will be described.
- *Cell therapy and disease's diagnosis:* In the field of cancer treatment, cell therapy has gained a lot of attention as a viable alternative to classical chemotherapy, in which side effects, especially the damage of healthy cells or even the emotional affectations, is still undesired [408]. In this sense, smart polymeric nanoparticles have opened the possibility of using local chemotherapy, acting directly in the damaged cells. Furthermore, another interesting approach is related to cancer immunotherapy, in which the main objective is to enhance the immune response of our system to fight cancer spreading, avoiding the use of cytotoxic drugs [409]. Lastly, we will describe the relevance of smart polymers in the diagnosis and treatment of Alzheimer, one of the most important degenerative diseases, affecting up to 50 million people worldwide in 2016 [410].

4.4.2 Precision medicine

The role of smart polymers in different aspects of precision medicine will be analysed in three medical procedures. First, non-invasive surgery, focused on minimizing the damage in body tissues when surgical procedures are carried out. Second, the innovative use of bio-inks in 3D bioprinting processes will be described. Finally, their use in cancer research applications defines the key points of microfluidic biomedical devices based on smart polymers and their importance in recent biomedical applications.

4.4.2.1 Non-invasive surgery

Non-invasive (or minimally invasive surgery, MIS) looks for the implantation of medical devices in an easy way, or if possible, removing these devices after the surgical procedure, minimizing the risks in the patient's body, reducing trauma and accelerating the recovery time. The main family of smart polymers which play a key role in these applications are the shape-memory polymers (SMPs), in which the external stimuli induce a controlled shape change of the smart polymer. Normally, the external stimulus is the temperature (heat-activated) or pH variation (pH or solvent-responsive) materials.

SMPs return to their original shape after being highly deformed upon exposure to a stimulus such as temperature, light and pH. This behaviour is usually triggered by supramolecular interactions (un- and re-coupling of non-covalent interactions including hydrogen bonds, host–guest and metal–ligand interactions) or dynamic covalent bonds (dissociation and recombination including boronate ester bond, imine and disulphide bonds). They are excellent candidates for stents, artificial muscles and MIS. Shape memory effect allows the shrinking of surgical devices to a smaller size, and they prevent tissue infection and in-growth, and thrombosis risk associated to drug eluting stents is reduced.

Among the vast family of shape-responsive materials, polyurethane foams have been employed since the early 2000s to design MIS devices [411], in specific applications such as the removal of clots in vascular thrombus, using photothermal activated SMP devices capable of trapping the clot through mechanical shrinking and subsequent removal after expansion [412], or employing these smart polymers as alternative aneurysm occluding materials [413]. Recent advances are focused on the use of biodegradable SMPs, to avoid a secondary surgery to retrieve the device. The most important point concerning the design requirements is the mechanical properties, such as stiffness or Young's modulus. For example, using SMP as vascular stents requires optimal rubbery Young's modulus between 1 and 10 MPa (low values can cause the stent to deform, and high values will difficult the stent recovery, or damage the arterial wall) [414]. The mechanical properties of SMP materials can be tuned through the cross-linking density, which is easily tailored during the curing process of the material.

When heat-activated SMPs are used, the glass transition temperature has a predominant influence on the shape recovery speed, and glass transition temperatures close to the body temperature are desirable. On the other hand, if the device is based on pH or solvent-activated SMP are employed, diffusion of body fluid into the inner surface of the material depends specifically on the thickness, and can take from minutes to hours [415].

Biocompatibility is an essential characteristic in medical applications. In this sense, smart polymers such as aliphatic polyesters, PLA, PCL or PEG can be used in many implant biomaterials and therapeutic devices, and most of them have been already approved by the FDA [416, 417]. Also, if the biomedical device is directly in contact with blood (e.g. vascular stents), hemocompatibility interaction tests is compulsory [418]. On the contrary, for non-permanent biomedical devices, biode-gradable polymers can carry out their functions for a given period of time and then degrade into small molecules, which can be eliminated from human body without any medical procedure [419]. In this line, PCL and PEG copolymers derivatives pres-ent a good biodegradability behaviour with weight losses up to 50% after 8 weeks at human body conditions [420, 421].

To conclude, among the many different polymer processing methods to pro-duce SMP materials, traditional casting and moulding procedures has been em-ployed classically, but the development of easy 3D printing technologies in the last years (in which biocompatible materials can be easily integrated) has become the predominant research line around biomedical devices, and for this reason we will dedicate the next section to this novel topic [422].

Nowadays, there are different applications related to MIS in which smart poly-mers have gained importance. In the last years, numerous researchers have demon-strated several possibilities of using SMPs as biomedical devices [423]. For example, poly(ethyleneglycol dimethacrylate) based SMP can be used as orthopaedic medical device for labrum injuries in the shoulder, adjusting the mobility above or below the glass transition temperature of the smart polymer [424]. Biodegradable thermo-plastics SMPs fibres are also employed as sutures for wound closure, as showed in the work of Lendlein and Langer [425]. However, the most representative SMP-based implant biomedical applications in which MIS is the key factor are vascular stents and occlusion devices, the former used mainly in heart attacks, and the latter to avoid aneurysms, and so they are discussed more extensively.

Heart attack is one of the most important causes of death around the world, and treating coronary artery diseases using stents has become of great importance. Most of the stents employed in current clinics are made of non-degradable materi-als, remaining in the coronary vessel, which could result in subsequent target-lesion failure or the apparition of stent thrombus [426]. For this reason, the use of biodegradable stents based on smart polymers could overcome these problems. In this line, there are several recent works using thermo-activated SMPs, with a trig-gering temperature close to the body temperature. Another important side advan-tage is that using these SMPs-based stents, specific drugs can be delivered in a controlled way through the mechanical response of the material, with drug libera-tion rates which can slowly increase for periods up to 30 days [427]. Acrylate-based amorphous SMPs with their onset glass transition temperature around body temper-ature can be programmed and stored in a catheter. Then, by putting them in contact with water at 37 °C, the SMP can recover the full size and shape in around 100 s [414]. Another type of biodevices which can be used as permanent stents are based in polymer blends of PU and PCL, which can recover their original shape (spring-like stents) at around body temperature [428]. Concerning the use of biodegradable smart polymers, Chen *et al.* fabricated chitosan-based stents, activated for shape recovery in aqueous media. These materials were completely biocompatible, biodegradable and non-toxic, thus showing great potential for non-permanent stent devices [429]. On the other hand, there are still some limitations about using biodegradable polymers, mainly focused on their low mechanical properties which could limit their performance and effectiveness [430].

Occlusion devices are intended to avoid the appearance of aneurysms, a very risky disease which eventually can lead to mental disabilities or death. The classical treatment method uses endovascular emboli to induce thrombosis, avoiding the rupture of the aneurysm [431]. SMP foams are currently investigated as a good candidate to fabricate occlusion devices, because of their excellent biocompatibility and biodegradability. Also, the infiltration of cellular components in the polymeric foam through the porous network can cause a larger fraction of volume occlusion, a very difficult achievement when classical metallic devices are employed [432].

4.4.2.2 3D bioprinting processes in medical applications

The emerging new technologies related to 3D printing have also been applied into the area of medical applications, opening a new research investigation field called "3D bioprinting" medical devices, which has even nowadays given a step forward to 4D bioprinting materials, focused specifically on SMPs [433]. In general, bioprinting is a promising and innovative fabrication technique to obtain, in a simple way using cells together with other stimuli-responsive materials such as hydrogels, different medical functional tissues and organs with complex geometries based on smart polymers that can be employed *in situ*, taking advantage of the triggered properties of these materials, and more specifically, their shape-memory behaviour [434].

In this context, smart polymers together with different cells have become the named "bio-inks", or the starting material which is impressed using an appropriate equipment. A simple example is shown in Figure 4.17, in which PEG-based hydrogels are employed as bio-inks. Using these materials, bi-layered constructs made of photo-cross-linkable polyethylene glycol with different molecular weights is presented (Figure 4.17a). In an aqueous environment, pre-designed folding of the constructs was induced by the differential swelling of the bonded hydrogel layers (Figure 4.17a), resulting in a well-predictable and controllable shapes: micropatterned spheres (Figure 4.17b), helices (Figure 4.17c) or cylinders (Figure 4.17d).

The polymers are initially in solution form, and then loaded with the desired cell types before the fabrication of the biomedical device. Different strategies are followed to fabricate these devices, taken from the classical polymer science technology: inkjet-based technologies, laser-assisted methods or extrusion technologies, although new fabrication methods are being now implemented, such as acoustic bioprinting or stereolithography bioprinting [435, 436]. Regarding the polymers used in bio-inks, natural, synthetic or a combination of both polymers can be used.

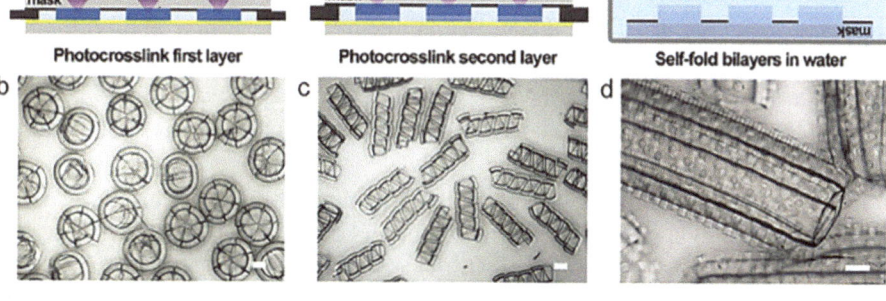

Figure 4.17: Fabrication of different shape-memory materials based on PEG hydrogels: (a) bi-layered constructs from photo-cross-linked PEG; (b) micropatterned spheres; (c) helices; (d) cylinders. Reproduced with permission of Elsevier [434].

Table 4.14 presents some examples of polymers employed as bio-inks and different references where more detailed information can be found.

Table 4.14: Some examples of smart polymers employed in bio-inks (more detailed information can be found in the book published by Matai *et al.* [437]).

Polymer type	Polymer	Reference
Natural	Collagen, silk, agarose, hyaluronic acid (HA)	[438, 439]
Synthetic	PEG, methacrylated hyaluronic acid (HAMA) Poly(*N*-vinyl-2-pyrrolidone) (PVP), poly(µ-caprolactone) (PLC), poly (lactic acid)	[440, 441]
Combination	Agarose/chitosan, PEG/HA, PLA/PEG	[442, 443]

The ability of bioprinting to simulate the human environment makes them suitable for different biomedical applications, from the classical fabrication of biomedical devices to more recent studies around cancer research treatments. In this sense, fabrication of 3D cancer models via bioprinting allows to improve the understanding of cancer behaviour and metastasis advancement, in order to develop advanced cancer therapies [444]. Inkjet bioprinting technique has shown to be able to print human ovarian cancer and normal fibroblasts cell lines, enabling the fabrication of cancer co-culture models for a better investigation of the cancer evolution [445]. This novel research field is evolving quickly, in order to simulate the behaviour of tumour cells using *in vivo* models to enhance and boost the diagnosis and treatments of many cancer types [446].

4.4.2.3 Microfluidic devices based on smart polymers

In recent years, microfluidic devices have found an interesting application field in the area of biomedicine [447]. Under the term "microfluidics", it is described the control and manipulation of fluids at sub-millimetric scale, which enables them to provide a biomimetic environment for cell and tissue culture [448]. Usually, different substrates such as glass, polymers and specially hydrogels are employed to fabricate the microfluidic device, which is composed of complex microchannels ranging from 1 to 500 μm [449].

Practical applications of microfluidic technology in biomedical applications using smart hydrogels include diagnosis chips and drug delivery devices, taking advantage of the accurate control of the transport of biomolecules [450]. The control and regulation of the biomolecule flow is carried out through microvalves and micropumps, and they are the main elements in which smart polymers can play a crucial role due to their responsive characteristics [451].

Microvalves are essential components for microfluidic systems, and they are employed as turn-on turn-off switches to regulate the flow of liquids at certain signals. Placing a stimuli-responsive hydrogel inside a microfluidic channel opens or closes the fluid flow, without needing external power sources, using only the triggering property of the smart hydrogel. Following this line, a SMP based on PLC and poly(dimethyl siloxane) (PDMS) was successfully employed as a heat-sensitive microvalve, as presented schematically in Figure 4.18 [452]. The microfluidic device is built on a glass substrate, and the microchannel is open or closed through the PCL shrinking behaviour, which is thermally controlled through the microheater placed in the bottom of the device. Other examples use photopolymerizable polymers as responsive materials, to fabricate a simple check valve whose operation mimics that of venous valves, based on a PDMS platform using 4-hydroxybutyl acrylate as elastic and photopolymerizable material [453]. More recently, a very interesting application of this technology was presented using thermo-responsive PNiPAAm microvalves (with diameters up to 600 μm), placed into PDMS-on-glass to obtain microfluidic devices for the control of parallelized enzyme-catalysed cascade reactions [454].

The other important component in microfluidic devices are micropumps, which promote fluid flow. In this case, fluid flow is triggered through the volume expansion of smart hydrogels. Different examples can be found in the literature. A classic work of Agarwal et al. is based on the use of thermo-responsive poly(HEMA-co-DMAEMA) hydrogels which started pumping fluids only above 40 °C [455]. A direct biomedical application was presented by Hara and Yoshida, who synthetized quaternary copolymers containing methacrylamidopropyltrimethylammonium chloride which could control their volume shifts through a redox reaction (Belousov–Zhabotinsky reaction) in the presence of malonic acid. The hydrogel could be induced to a self-oscillating movement, mimicking the behaviour of an artificial muscle [456]. Recent studies have shown the potential use of these devices as advanced hand-held

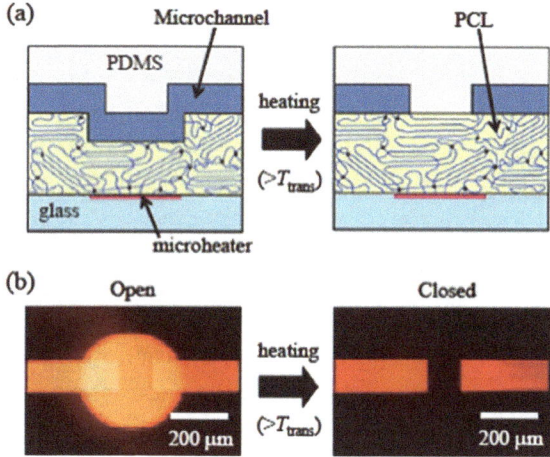

Figure 4.18: Example of SMP-based microvalve: (a) scheme of a microfluidic device; (b) micrographs of SMP open and closed states. Reproduced with permission of J-Stage [452].

drug delivery devices, fabricating leaf-inspired micropumps based on agarose hydrogels, with the ability of providing a durable flow rate of 30 µL/h · cm^2 for more than 30 h at RT without external mechanical power source [457]. To conclude, there also a few studies analysing the opposite behaviour, using smart hydrogels in microfluidic suction pumps, which could be applied, for example, in cleaning and removal processes of the suppuration fluids derived from chronic wounds [458].

4.4.3 Cell therapy

Cell therapy is aimed to treat specifically a cell or group of cells using smart polymeric nanoparticles able to deliver drugs in a controlled way into a targeted site, without affecting the surrounding healthy cells. In this last part of the chapter, the basis of the cell therapy using smart polymers will be stablished. As drug delivery polymers have already been discussed in **Section 4.2**, this chapter will be focused just in concrete medical applications: cancer treatment (Section 4.3.1), and the analysis of the recent advances in the therapeutic role of smart polymers in degenerative diseases, such as Alzheimer or Parkinson.

4.4.3.1 Cancer treatment
Common cancer treatment is based on chemotherapy [459], which, however, presents many drawbacks in the drug delivery procedure, such as non-specificity, low drug concentration in tumour tissue and systemic toxicity, leading to damage the tumour-bearing host and its immune system. Then, different biological alternatives, such as

immunotherapy, have emerged in recent years as a promising technique to limit these side effects [407].

The use of smart polymeric hydrogels in cancer therapy is carried out in three main scenarios. First, due to the difference between the pH observed in healthy and tumour cells, it is possible to use pH-responsive polymeric hydrogels for controlled and targeted release of specific drugs due to the abnormally low pH of endosomes and tumour cells compared to healthy ones [460]. Second, redox-responsive polymers can also be employed for drug delivery systems, using essentially disulphide linkers, which are reduction-sensitive and can be cleaved specifically at high concentrations of GSH, a tri-peptide consisting of glutamate, cysteine and glycine. It has been observed that tumour tissues present higher GSH concentration that healthy tissues, then opening the possibility of using the GSH concentration as a redox trigger for controlled drug delivery [461]. Last, it is also possible to attack specific receptors that can be found exclusively in cancer cells, such as folate, and release their drug directly [462].

In all the cases, due to the exceptionally low sizes required for these applications, it is necessary to employ small smart nanoparticles (polymeric micelles). These are able to encapsulate hydrophobic drugs in their core due to their amphiphilic nature, assembling in similar patterns to the ones observed in viruses and lipoproteins [463]. The main advantage of smart polymeric micelles is that they can lower the dosage frequency, then retaining the drug concentration in the targeted site for longer periods. On the other hand, although many different studies have been carried out to stablish these materials as drug carriers, only a few of them have reached into clinical studies [464], mainly due to the need of combined high biocompatibility and biodegradability.

a) pH-Responsive polymers in tumour treatment

Tumour micro-environment is ideal for triggering the drug release within tumour cells. Extracellular regions in normal tissues and blood have a pH around 7.4, but extracellular regions of tumour cells have a pH considerably lower, between 6.0 and 6.5 [23]. pH-sensitive polymeric micelles have been used for targeting drug delivery to tumours due to their stability at physiological pH, and they can be also deformed physically to enhance the release the drug under mildly acidic conditions outside the tumour cells, as it is shown in Figure 4.19. In this case, the anticancer drug is encapsulated into the polymeric micelles via acid-labile bonds, such as hydrazone, *cis*-acotinyl or acetal.

In the process showed above, drug release is controlled through the cleavage or rupture of the acid-labile bonds. In these polymeric micelles, the presence of acid-sensitive spacers between the polymer and the drug facilitates the drug release when pH is lowered in the tumour tissues region. Other strategy than can be followed to tune the drug release is the use of smart polyelectrolytes, which can be

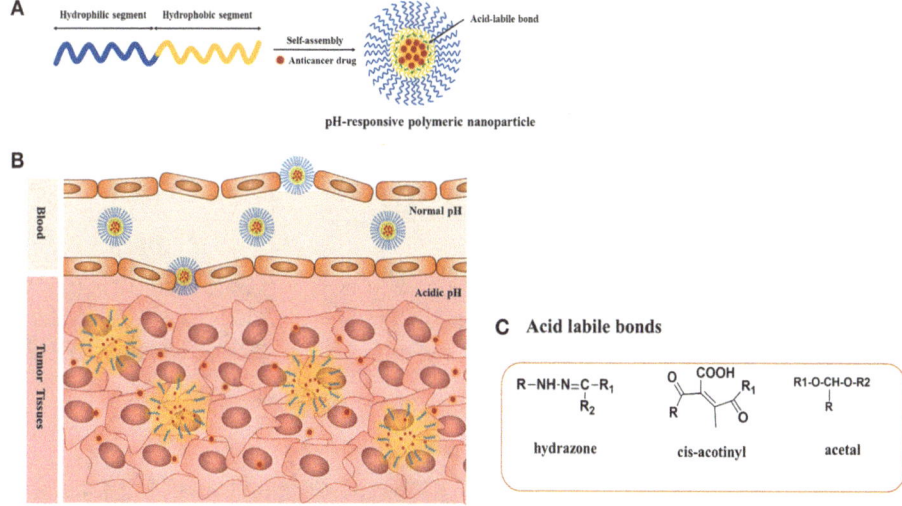

Figure 4.19: Schematic behaviour of pH-responsive polymeric micelles; (a) design and encapsulation of drugs into the polymeric micelle; (b) drug release mechanism under the acidic environment such as solid tumours, endosomes or lysosomes through the cleavage of the acid-labile bond; (c) chemical structures of some chemical bonds. Reproduced with permission of Frontiers Media SA [465].

ionized by changing the environmental conditions or the pH, releasing the drug at the same time [196].

Different examples can be found in the literature using the procedure exposed in Figure 4.20. Polymeric micelles based on PLA and PEG in which DOX was encapsulated show good biocompatibility and biodegradability, and the drug is released in extracellular tumour environments showing acidic media (pH around 6.2) [466]. Also, polymeric micelles based on poly(N-vinyl-2-pyrrolidone) and dimethyl maleic anhydride encapsulating DOX through an acid-sensitive linker, are able to release the drug when pH was reduced in contact to tumour cells [467]. In most of the cases, DOX has shown a good ability to be encapsulated in pH-sensitive polymeric micelles, and for this reason, the number of research works in which DOX is employed has increased greatly. The main problem associated to the use of DOX as drug is that, although this pH-sensitive micelles can reduce the systemic toxicity, the issue of multidrug resistance of tumour cells to anticancer drugs is still unsolved. Then, it is necessary an initial rapid release process of the drug from the micelles into the tumour cells, and the fastening of the triggering process is still under investigation. In this line, a recent work employed five types of micelles made of poly(2-oxazoline) block copolymers in which dexamethasone was encapsulated. Then, ultrasound was combined to pH stimulus as triggering unloading mechanism to increase the initial amount of released dexamethasone up to ten

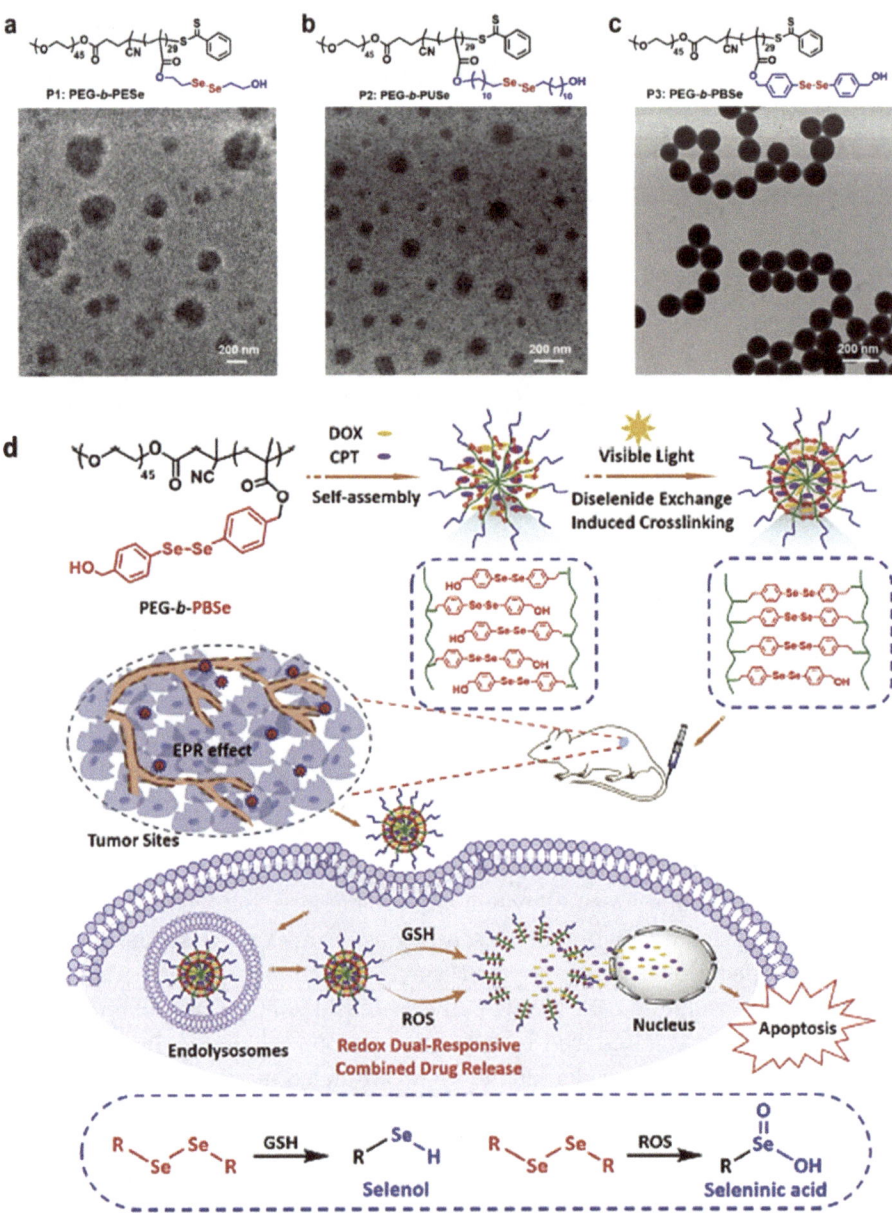

Figure 4.20: Using diselenide-bond based micelles for encapsulation and drug release in tumour cells: (a), (b) and (c) chemical structures and TEM micrographs of the smart-polymeric micelles synthetized; (d) scheme of the encapsulation and drug release into tumour sites process via redox-responsive mechanisms (EPR indicates the "enhanced permeability and retention effect" and ROS are "reactive oxygen species", which have been found at elevated concentrations in tumour cells, leading to the appearance of redox dual-responsive effect in drug release). Reproduced with permission of Elsevier [471].

times, depending on the type of copolymer [468]. Other anticancer drug that can be used in pH-responsive micelles is paclitaxel, which is one of the most effective chemotherapeutic drugs ever developed, active against a broad range of cancers (lung, ovarian or breast). This anticancer drug can be easily encapsulated in biodegradable and non-toxic micelles based on different polymers, such as PLA, PLG or chitosan. In addition, paclitaxel albumin-bound nanoparticles (commercial brand Abraxane®) were approved by the FDA in 2013 for the treatment of metastatic breast cancer and non-small cell lung cancer [469].

b) Redox-responsive polymers

Redox-responsive polymers have been employed classically for disease therapy and controlled drug release applications, although in recent years, great process has been developed in this research field [220]. The reducing environment of tumours is a unique internal signal to degrade selectively redox-responsive nanocarrier and release the specific drugs. These systems offer different advantages: stability in healthy tissues (reducing the toxicity and side effects), fast response to the glutathione GSH concentration in tumour cells to rapidly release the drugs (starting from a few minutes) and finally the drug release in cytoplasm can have also therapeutic effects [470]. The most studied stimuli-responsive polymers employed as redox-responsive drug releasers are the ones containing disulphide bonds (see **Section 4.2.** for further information and examples of the type of polymers used). At a minor extent, diselenide bonds are being employed, due to their similar reduction sensitivity and redox-responsive ability to the one observed in disulphide bonds. Also, Se–Se and C–Se bonds present lower bond energies than that S–S bonds (Se–Se 172 kJ/mol; C–Se 244 kJ/mol; S–S 268 kJ/mol), which allows the design of more sensitive systems using these smart polymers [471]. The use of disulphide-based smart polymers is focused on the fabrication of self-assembled amphiphilic copolymers via disulphide-containing cross-linkers, oxidation of thiol groups and disulphide–thiol exchange reaction. These disulphide bonds are included in the backbone of the polymer or, in a different approximation, a GHS-sensitive cross-linking agent is used either in the shell or in the core of the micelles.

The cleavage of the disulphide bond is the main mechanism to release the drug into the tumoural cell. Different research works employ these types of encapsulation micelles in cancer therapy. For example, redox-responsive tumour-targeted micelles based on HA and PCL can be used through the disulphide bonding of the HA shell and subsequent cleavage, releasing DOX with high loading efficiency [472]. A scheme showing the redox-responsive smart polymer role in the tumour-targeted drug delivery is presented in Figure 4.7 in **Section 4.2.**

An alternative is the use of diselenide-rich amphiphilic diblock copolymers [471]. In this case, two different drugs, camptothecin and DOX, with synergistic antitumour properties are encapsulated in cross-linked micelles based on PEG and polybutylene succinate (PBS). Cross-linked process was easily carried out using visible light, obtaining the micelles presented in Figure 4.20a, 4.20b and 4.20c, in which the Se-Se bonds are placed at the backbone of the main chains for the three micelles synthetized with distinct hydrophobic units. The schematic illustration for the self-assembly and visible light-induced core cross-linking of diselenide-rich nanoparticles fabricated from PEG–PBS diblock copolymers is presented in Figure 4.20d.

c) Targeting folate using smart polymers

An interesting alternative to use the therapeutic potential of smart polymers in cancer therapy is the targeting of specific receptors that are found exclusively on cancer cells, such as folate.

Folate receptor is abundant in many cancer cell types, such as kidney, lung, breast, bladder and pancreas [473]. Then, folic acid, which can recognize specific receptors such as the overexpressed folate, can be used in pH-responsive polymeric micelles to interact specifically with folate via ligand-receptor interactions, triggering the drug release. Using this mechanism, poly(styrene-alt-maleic anhydride) amphiphilic polymer can be modified and functionalized with the addition of folic acid. The obtained cylindrical micelle not only maintains its pH responsiveness to release controllably the therapeutic drug, but in addition, it can be bonded to the folate receptors present on cancer cells [474]. The intact cylinder can travel through the body, bind to the folate receptor and then release the therapeutic drug inside the tumour cell due to the conformational change driven by the pH drop. Using these polymeric micelles, curcumin has been successfully loaded and encapsulated, employing this molecule in cancer therapy due to its anti-inflammatory properties [475]. In another example, curcumin can be loaded onto folate targeted Au-poly(N-vinyl-2-pyrrolidone) nanoparticles, showing up to 80% release at acidic environments, specifically to treat breast cancer, obtaining very promising results [476]. The importance of these functionalized micelles with folic acid is that it can be used to encapsulate both synthetic drugs, such 5-fluorouracil or paclitaxel, and also natural drugs, such as the mentioned curcumin, to compare their therapeutic behaviour in different cancer types [474]. Figure 4.21 presents the chemical structure of these drugs, commonly used in cancer therapy.

Although there are many research works involved in these therapy strategies, some difficulties must be still overcome to assure the efficacy of smart polymers in human cancer treatment. In this sense, animal models (mouse or rats) are indeed tools for understanding the mechanisms and molecular basis of pathology diseases, but these approaches cannot accurately reproduce human diseases, then limiting the clinical application. For this reason, different alternatives, such immunotherapies, are gaining importance in recent years, due to the absence of cytotoxic drugs and, subsequently, all their side effects.

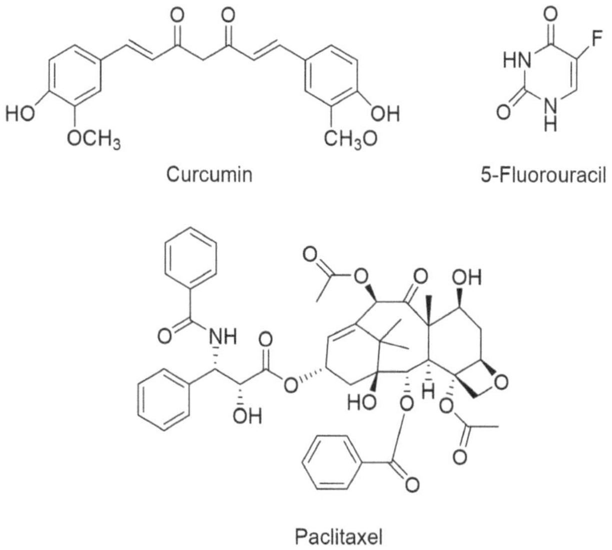

Curcumin   5-Fluorouracil

Paclitaxel

Figure 4.21: Chemical structure of therapeutic drugs employed in cancer therapy.

There are different approaches for immunotherapies, such as cancer vaccination, immune checkpoint therapy or adoptive cell transfer. However, conventional immunotherapies present immune-related side effects, such as liver dysfunction or diabetes. For this reason, there are new and promising alternatives in which smart polymeric nanoparticles could address these issues [407]. The most advanced research field is related to "cancer vaccine", based on the controlled release of antigens into the human body to promote and enhance the immune response. In this sense, polymeric micelles are employed in antigen delivery, targeting and activating the antigen-presenting cells to enhance innate and adaptive immunity [477]. For example, pH-sensitive smart polymers based on decalysine-modified antigen peptides and poly(propylacrylic acid) can be used to enhance antigen presentation, controlling the antigen release profile facilitating a persistent immune response [478]. To assure successful vaccination, the administration route is a key factor, that can be optimized using intradermal nanovaccines implemented using microneedles, a strategy which has emerged in the last years to optimize the efficiency of the immunotherapy [479]. Different examples can be found in the literature describing the role of polymeric micelles in cancer vaccination. Dissolvable microneedles arrays loaded with antigen encapsulated micelles based on poly(lactic-*co*-glycolic acid) can be obtained to assure controlled and sustained release of antigen in the skin. After dissolution of the microneedles in the skin within minutes, antigen-loaded micelles are exposed to the skin layer, releasing the antigen in a controlled way [480]. In a similar way, a smart DNA vaccine delivery system was developed using microneedles assembled with layer-by-layer coating of pH-responsive smart polymers based on poly(β-amino ester urethane) and sulphamethazine, finding that a controlled release of vaccines and adjuvants was achievable at the immune cell-rich epidermis/dermis layers, using a skin implanted layer-by-layer coated microneedle strong enough to traverse the *stratum corneum* [481]. As it can be seen, the number of research works and different alternatives about the use of smart polymers in cancer therapy is large and it will continue to grow in the coming years, pointing out the importance of these materials in this crucial investigation field.

4.4.3.2 Recent advancements in nanotherapeutics of degenerative diseases

Neurodegenerative diseases, such as Alzheimer or Parkinson, are caused by the progressive dysfunction and death of neurones. These diseases are becoming strongly prevalent, and their impact is increasing with advancement of age. These neurodegenerative disorders are due to complex formation process of different cytotoxic protein aggregates, causing neurodegenerative stressing in the brain leading to oxidative and inflammatory problems in the CNS [482]. The special environment in which neurodegenerative diseases take place, the CNS, is responsible of some specific limitations that bio-therapeutic drugs must overcome to achieve good effectiveness [483]. CNS is double protected by two biological barriers: blood–brain barrier (BBB) and blood–cerebrospinal fluid barrier (BCSFB). Both barriers prevent the penetration of almost 100% of large molecules (more than 500 Da) and 90% of smaller drug molecules [484]. There is also a third negative effect caused by the activity of P-glycoprotein (also known and multidrug resistance protein, which acts as drug efflux limiting the effectiveness of a wide spectrum of therapeutic drugs as well as chemotherapy). For all these reasons, the use of smart nanomaterials, and specifically, smart polymeric micelles, has become very attractive in the last years, also taking advantage of their large surface area to volume ratio. It has been demonstrated that smart polymeric lipophilic micelles with sizes of 100 nm or lower can pass easily through the BBB barrier through diffusion mechanisms, releasing the therapeutic drug into the CNS [485].

Due to the need of using these polymeric micelles inside the CNS, it is essential to assure their biodegradability or at least they must be easy to remove completely and quickly from the body. Also, they must be non-toxic and their decomposition products must not present immunogenic side effects. There are recent studies showing that DDS have successfully penetrated the BBB barrier to deliver the cargo drug.

Nowadays, there are different drugs approved by the EMA, the FDA or other governmental agencies to treat these diseases which are administered orally: memantine, donepezil, rivastigmine and galantamine. These drugs are not curative, and they can only treat different symptoms of the diseases, increasing the quality life reducing the cognitive loss and enhancing motor regulation.

Also, they present different problems. First, the release rate of these oral drugs is not well controlled, leading to quick release rates which limit the therapeutic effect to a short period of time. Also, as commented before, the presence of the BBB and BCSFB limit greatly the effectivity of these oral drugs. Then, smart polymeric micelles are designed to overcome these limitations through three different key points [486]:

- Reduction of the scale of the carrier to nanometre size to penetrate into the CNS through diffusion mechanisms in BBB and BCSFB barriers.
- Protection of hydrophobic drugs from degradation and deactivation, increasing their availability.
- Use of stimuli-responsive polymers to trigger the drug release and optimize their effectivity in longer time periods (up to 14 days).

Literature present very promising works in which smart polymeric micelles are used are being employed to treat these neurodegenerative diseases. For example, the synthesis of polymeric micelles based on *n*-butyl-2-cyanocrylate was investigated to be used for Alzheimer treatment. In a second step, the drug clioquinol was successfully encapsulated. These micelles were loaded into a mouse model and were able to penetrate the BBB barrier, significantly increasing the brain retention of the drug [487]. Natural smart polymers, such as chitosan, were used also in combination with rivastigmine to treat Alzheimer disease. Polymeric micelles of 47 nm were synthetized to encapsulate this drug by emulsification, observing a well-controlled and triggered pH release rate (following a Fick's diffusion profile) enhancing drastically its therapeutic effect [488]. Also, an alternative to oral drug administration for Alzheimer's disease was recently developed by synthetizing a smart polymer based delivery system for controlled release of rivastigmine using a simple subcutaneous injection. The polymeric micelle was fabricated from poly (lactic-*co*-glycolic acid) (50:50) at 5% w/v in benzyl benzoate. The benzoic acid demonstrated excellent rivastigmine drug release characteristics *in vitro* up to 98 days, with a sustainable drug profile over time, good biocompatibility and good ability to penetrate BBB barrier after injection. The use of these injectable smart polymeric micelles opens a new research field focused on the development of more efficient treatments of neurodegenerative diseases [489].

5 Conclusions, challenges and perspectives

The concept of smart polymer is wide and somewhat vague, and it is difficult to pack it in a single box or even set of boxes with a precise meaning, constitution or even objective. Within this context and related with the broad concept of human or artificial intelligence, smart polymers have been considered in this manuscript to be able to process signals from the environment, both to perform an action (responsive polymers) or to emit another signal that can be used as functional, analytical information (sensory polymers). And for both, the triggering stimuli may be physical or chemical, the latter including biological. In any case, what is common to any consideration, classification or analysis of these materials is its ultimate objective within the framework of the development of smart devices, actuators and sensors for advanced applications.

The detection and quantification of chemical species, in solid state, solution and as gases and vapours, is a traditional and relevant topic of analytical chemistry of obvious interest in the industry, chemical and safety control, in research laboratories, and in the environment, in pollution control in soils, water and air. In last decades, analysing these species using supramolecular chemistry, exploiting host–guest interaction that led to measurable macroscopic changes, has represented a breakthrough in terms of the ability to perform simply, easily, quickly and cheaply on-site analyses, giving rise to the chemosensors field. In a step forward, the development of polymers with guest moieties in their structure has given rise to sensory materials with the ability to respond to different stimuli, single or multiple, that can be transformed easily into different shapes, such as coatings, films, fibres or wires. At the same time, the hosts or sensory moieties chemically anchored to the macromolecular chains impair the migration of these moieties, increasing not only the security of the sensors for use in biomedical and food applications, but also allows for their reusability. Also, the chemical structure of the polymer main chain can be easily tuned to achieve the hydrophilicity degree needed for broadening the applicability of the sensory polymers for the detection of harmful substances (volatile organic compounds and explosives, heavy metal cations, anions of analytical interest, organic and inorganic molecules of industrial, biological of environmental interest and biomolecules in biological media). The challenge is the development of selective and sensitive polymers for preparing different sensory smart devices for the naked eye detection of species, for pairing with mobile applications for the fine quantification of chemicals, and for wearables related, for example, with smart fibres for intelligent textiles. The biological recognition phenomena are successfully exploited in biosensing with the immobilization of biomolecules, such as enzymes, antibodies and antigens for the recognition of peptide substrates and other proteins, including antigens and antibodies. The success of this strategy is based on the immobilization of these biomolecules, of a part of them, in a polymeric support.

https://doi.org/10.1515/9781501522468-005

This support is considered intelligent when it plays a significant role in the correct performance of the biosensor, usually related with the immobilized biomolecules, or rest, and specifically with their orientation that prevents denaturalization, access to the target species and a correct space interaction of target and receptor. Thus, the challenge, from the polymer viewpoint, is to provide protection to the bioreceptor, correct orientation in the immobilization and access of the target to the bioreceptor.

The precise control of the polymeric chemical structure, along with a deep knowledge in-between chemistry, biology and medicine, is a field of opportunities to address new challenges in medicine, and specifically in precision medicine centred on the individual. The potential use of smart polymers in biomedical applications has lately emerged representing the technology behind different applied areas related with drug delivery, and especially anti-cancer drug delivery, insulin delivery, gene delivery and so on, in oral, topical and parenteral delivering systems based on hydrogels or nanostructures (nanoparticles and coated nanoparticles, and nanofibers). Despite the huge progress regarding these smart materials, some challenges still must be addressed to be used in human clinical trials, such as biodegradability and biocompatibility. Most of the developed materials have only been used *in vitro* as a proof of concept, and many extra efforts must be made regarding the loading capacity, programmable release, *in vivo* stability, non-toxicity, biodegradability and so on, to be used in biomedical applications. A relevant application in biomedicine is tissue engineering, related with the ability of synthetic biopolymers to mimic natural materials for maintenance, or even restoration or reconstruction of a complete organ. The rolls of the polymers are to serve as scaffold materials acting as architecture for adhesion, transplantation and cell function, bioreactors as active media for cell proliferation, and signalling molecules to bind receptors. The challenge is to go further providing not only mechanical support but also biomimetic characteristics with enhanced interaction with human tissue, thus designing biopolymers as hydrogels responding to stimuli, such as pH, temperatures and magnetic field, for controlling cell attachment and growth, cell-cell communication and proliferation and differentiation. Also, the challenge is related with exploiting in 3D printing and direct injection of smart polymers capable of recovery or change the shape (self-healing and shape-memory) or respond to temperature, pH and so on, for cell, drug and protein carriers and proliferation. Related with this challenge, precision medicine is emerging integrating several fields, such as smart biomaterials, medical and sensory devices, large-scale databases and data mining and medical specialties. Similarly, the customization of smart biopolymers, stem cells, bioprinting and transplantation has emerged as a new field of opportunities for disease treatment by cell therapy, where polymers may have a major role regulating the differentiation of stem cells under external stimuli. However, in biomedicine, the polymeric structure, usually considered inert, is worth to be exploited as smart materials with biological activity as antibiotics, antiviral, antitumour, antithrombotic and so on.

References

[1] Berzelius J. Isomerie, Unterscheidung von damit analogen Verhältnissen. Jahres-Bericht. 1833;12:63–7.

[2] Staudinger H. Über Polymerisation. Berichte der Dtsch Chem Gesellschaft (A B Ser.) 1920;53 (6):1073–85.

[3] Staudinger H, Fritschi J. Über Isopren und Kautschuk. 5. Mitteilung. Über die Hydrierung des Kautschuks und über seine Konstitution. Helv Chim Acta. 1922;5(5):785–806.

[4] Braunecker WA, Matyjaszewski K. Controlled/living radical polymerization: Features, developments, and perspectives. Prog Polym Sci. 2007;32(1):93–146.

[5] Yokozawa T, Yokoyama A. Chain-growth polycondensation: The living polymerization process in polycondensation. Prog Polym Sci. 2007;32(1):147–72.

[6] Wong AD, DeWit MA, Gillies ER. Amplified release through the stimulus triggered degradation of self-immolative oligomers, dendrimers, and linear polymers. Adv Drug Deliv Rev. 2012;64 (11):1031–45.

[7] Aguilar MR, San Román J, editors. Smart polymers and their applications. Second. Woodhead Publishing; 2019.

[8] Sanjuán AM, Reglero Ruiz JA, García FC, García JM. Recent developments in sensing devices based on polymeric systems. React Funct Polym. 2018;133:103–25.

[9] Niskanen J, Tenhu H. How to manipulate the upper critical solution temperature (UCST)? Polym Chem. 2017;8(1):220–32.

[10] Seuring J, Agarwal S. Polymers with upper critical solution temperature in aqueous solution: unexpected properties from known building blocks. ACS Macro Lett. 2013;2(7):597–600.

[11] Gillies ER. Reflections on the evolution of smart polymers. Isr J Chem. 2020;60(1–2):75–85.

[12] Wang D, Wang X. Amphiphilic azo polymers: molecular engineering, self-assembly and photoresponsive properties. Prog Polym Sci. 2013;38(2):271–301.

[13] Vapaavuori J, Bazuin CG, Priimagi A. Supramolecular design principles for efficient photoresponsive polymer–azobenzene complexes. J Mater Chem C. 2018;6(9):2168–88.

[14] Yu H. Recent advances in photoresponsive liquid-crystalline polymers containing azobenzene chromophores. J Mater Chem C. 2014;2(17):3047–54.

[15] Bertrand O, Gohy J-F. Photo-responsive polymers: synthesis and applications. Polym Chem. 2017;8(1):52–73.

[16] Thévenot J, Oliveira H, Sandre O, Lecommandoux S. Magnetic responsive polymer composite materials. Chem Soc Rev. 2013;42(17):7099.

[17] Okuzaki H. Electroresponsive polymer. In: Kobayashi S, Müllen K, editors. Encyclopedia of Polymeric Nanomaterials. Berlin, Heidelberg: Springer-Verlag Berlin Heidelberg; 2014. p. 1–6.

[18] Brown CL, Craig SL. Molecular engineering of mechanophore activity for stress-responsive polymeric materials. Chem Sci. 2015;6(4):2158–65.

[19] Zhang H, Lin Y, Xu Y, Weng W. Mechanochemistry of topological complex polymer systems. In: Boulatov R, editor. Polymer mechanochemistry topics in current chemistry. Springer; 2014. p. 135–207.

[20] Ruiz JAR, Sanjuán AM, Vallejos S, García FC, García JM. Smart polymers in micro and nano sensory devices. Chemosensors. 2018;6:12.

[21] Colson YL, Grinstaff MW. Biologically responsive polymeric nanoparticles for drug delivery. Adv Mater. 2012;24(28):3878–86.

[22] Gillies ER, Fréchet JMJ. A new approach towards acid sensitive copolymer micelles for drug delivery. Chem Commun. 2003;1640–1.

[23] Kocak G, Tuncer C, Bütün V. PH-Responsive polymers. Polym Chem. 2017;8(1):144–76.

https://doi.org/10.1515/9781501522468-006

[24] Dai S, Ravi P, Tam KC. pH-Responsive polymers: synthesis, properties and applications. Soft Matter. 2008;4(3):435.

[25] Chen J-K, Hsieh C-Y, Huang C-F, Li P-M, Kuo S-W, Chang F-C. Using solvent immersion to fabricate variably patterned poly(methyl methacrylate) brushes on silicon surfaces. Macromolecules. 2008;41(22):8729–36.

[26] Chen J-K, Hsieh C-Y, Huang C-F, Li P. Characterization of patterned poly(methyl methacrylate) brushes under various structures upon solvent immersion. J Colloid Interface Sci. 2009;338 (2):428–34.

[27] Barsan MM, Ghica ME, Brett CMA. Electrochemical sensors and biosensors based on redox polymer/carbon nanotube modified electrodes: a review. Anal Chim Acta. 2015;881:1–23.

[28] Huo M, Yuan J, Tao L, Wei Y. Redox-responsive polymers for drug delivery: from molecular design to applications. Polym Chem. 2014;5(5):1519–28.

[29] Phillips DJ, Gibson MI. Degradable thermoresponsive polymers which display redox-responsive LCST Behaviour. Chem Commun. 2012;48(7):1054–6.

[30] Mazurowski M, Gallei M, Li J, Didzoleit H, Stühn B, Rehahn M. Redox-responsive polymer brushes grafted from polystyrene nanoparticles by means of surface initiated atom transfer radical polymerization. Macromolecules. 2012;45(22):8970–81.

[31] Peterson GI, Larsen MB, Boydston AJ. Controlled depolymerization: stimuli-responsive self-immolative polymers. Macromolecules. 2012;45(18):7317–28.

[32] Sirkar K, Revzin A, Pishko M V. Glucose and lactate biosensors based on redox polymer/oxidoreductase nanocomposite thin films. Anal Chem. 2000;72(13):2930–6.

[33] Pérez JPH, López-Cabarcos E, López-Ruiz B. The application of methacrylate-based polymers to enzyme biosensors. Biomol Eng. 2006;23(5):233–45.

[34] Liu X, Fan Q, Huang W. DNA biosensors based on water-soluble conjugated polymers. Biosens Bioelectron. 2011;26(5):2154–64.

[35] Hale PD, Boguslavsky LI, Inagaki T, Karan HI, Lee HS, Skotheim TA, et al. Amperometric glucose biosensors based on redox polymer-mediated electron transfer. Anal Chem. 1991; 63(7):677–82.

[36] Fleige E, Quadir MA, Haag R. Stimuli-responsive polymeric nanocarriers for the controlled transport of active compounds: concepts and applications. Adv Drug Deliv Rev. 2012; 64(9):866–84.

[37] Liu D, Yang F, Xiong F, Gu N. The smart drug delivery system and its clinical potential. Theranostics. 2016;6(9):1306–23.

[38] Qiu Y, Park K. Environment-sensitive hydrogels for drug delivery. Vol. 53, Advanced Drug Delivery Reviews. Elsevier; 2001. p. 321–39.

[39] Roy D, Cambre JN, Sumerlin BS. Future perspectives and recent advances in stimuli-responsive materials. Prog Polym Sci. 2010;35(1–2):278–301.

[40] Gunatillake P. Biodegradable synthetic polymers for tissue engineering. Eur Cells Mater. 2003;5:1–16.

[41] Rezwan K, Chen QZ, Blaker JJ, Boccaccini AR. Biodegradable and bioactive porous polymer/inorganic composite scaffolds for bone tissue engineering. Biomaterials. 2006;27(18): 3413–31.

[42] Huang H-J, Tsai Y-L, Lin S-H, Hsu S. Smart polymers for cell therapy and precision medicine. J Biomed Sci. 2019;26(1):73.

[43] Wiktorowicz S, Tenhu H, Aseyev V. Multi-stimuli-responsive polymers based on calix[4] arenes and dibenzo-18-crown-6-ethers. In: Khutoryanskiy V V., Georgiou TK, editors. Temperature-responsive polymers. Chichester, UK: John Wiley & Sons Ltd; 2018. p. 145–74.

[44] Tang X, Liang X, Gao L, Fan X, Zhou Q. Water-soluble triply-responsive homopolymers of *N,N* -dimethylaminoethyl methacrylate with a terminal azobenzene moiety. J Polym Sci A Polym Chem. 2010;48(12):2564–70.

[45] Seeboth A, Lötzsch D, Ruhmann R, Muehling O. Thermochromic polymers—function by design. Chem Rev. 2014;114(5):3037–68.

[46] Persaud KC. Polymers for chemical sensing. Mater Today. 2005;8(4):38–44.

[47] Wu Y, Dong Y, Li J, Huang X, Cheng Y, Zhu C. A highly selective and sensitive polymer-based fluorescence sensor for Hg^{2+}-ion detection via click reaction. Chem - An Asian J. 2011; 6(10):2725–9.

[48] Sheng Dai, Palaniswamy Ravi, Chiu Tam K. pH-Responsive polymers: synthesis, properties and applications. Soft Matter. 2008;4(3):435–49.

[49] Guembe-García M, Santaolalla-García V, Moradillo-Renuncio N, Ibeas S, Reglero JA, García FC, et al. Monitoring of the evolution of human chronic wounds using a ninhydrin-based sensory polymer and a smartphone. Sens Actuator B. 2021;335:129688.

[50] Pablos JL, Vallejos S, Muñoz AA, Rojo MJ, Serna F, García FC, et al. Solid polymer substrates and coated fibers containing 2,4,6-trinitrobenzene motifs as smart labels for the visual detection of biogenic amine vapors. Chem Eur J. 2015;21(24):8733–6.

[51] González-Ceballos L, Melero B, Trigo-López M, Vallejos S, Muñoz A, García FC, et al. Functional aromatic polyamides for the preparation of coated fibres as smart labels for the visual detection of biogenic amine vapours and fish spoilage. Sens Actuators B. 2020;304:127249.

[52] Trigo-López M, Muñoz A, Ibeas S, Serna F, García FC, García JM. Colorimetric detection and determination of Fe(III), Co(II), Cu(II) and Sn(II) in aqueous media by acrylic polymers with pendant terpyridine motifs. Sens Actuators B. 2016;226:118–26.

[53] Vallejos S, Reglero JA, García FC, García JM. Direct visual detection and quantification of mercury in fresh fish meat using facilely prepared polymeric sensory labels. J Mater Chem A. 2017;5(26):13710–6.

[54] Carvalho RN, Ceriani L, Ippolito A, Lettieri T. Development of the first watch list under the environmental quality standards directive water policy [Internet]. JRC technical report, Publications Office of the European Union 2015. https://publications.jrc.ec.europa.eu/reposi tory/handle/JRC95018.

[55] Heskins M, Guillet JE. Solution properties of poly(*N*-isopropylacrylamide). J Macromol Sci Chem. 1968;2(8):1441–55.

[56] Wu Y, Alici G, Madden JDW, Spinks GM, Wallace GG. Soft mechanical sensors through reverse actuation in polypyrrole. Adv Funct Mater. 2007;17(16):3216–22.

[57] Wang T, Farajollahi M, Choi YS, Lin IT, Marshall JE, Thompson NM, et al. Electroactive polymers for sensing. Interface Focus. 2016;6(4):20160026.

[58] Davis DA, Hamilton A, Yang J, Cremar LD, Gough D Van, Potisek SL, et al. Force-induced activation of covalent bonds in mechanoresponsive polymeric materials. Nature. 2009; 459(7243):68–72.

[59] Thévenot J, Oliveira H, Sandre O, Lecommandoux S. Magnetic responsive polymer composite materials. Chem Soc Rev. 2013;42(17):7099–116.

[60] Rather IA, Ali R. Indicator displacement assays: from concept to recent developments. Org Biomol Chem. 2021;19:5926–81.

[61] Sedgwick AC, Brewster JT, Wu T, Feng X, Bull SD, Qian X, et al. Indicator displacement assays (IDAs): the past, present and future. Chem Soc Rev. 2021;50(1):9–38.

[62] Nguyen BT, Anslyn E V. Indicator-displacement assays. Coord Chem Rev. 2006;250 (23–24):3118–27.

[63]   García-Calzón JA, Díaz-García ME. Characterization of binding sites in molecularly imprinted polymers. Sens Actuators B. 2007;123(2):1180–94.

[64]   Leibl N, Haupt K, Gonzato C, Duma L. Molecularly imprinted polymers for chemical sensing: a tutorial review. Chemosensors. 2021;9(6):123.

[65]   Vasapollo G, Sole R Del, Mergola L, Lazzoi MR, Scardino A, Scorrano S, et al. Molecularly imprinted polymers: present and future prospective. Int J Mol Sci. 2011;12(9):5908–45.

[66]   Haupt K, Mosbach K. Molecularly imprinted polymers and their use in biomimetic sensors. Chem Rev. 2000;100(7):2495–504.

[67]   McCluskey A, Holdsworth CI, Bowyer MC. Molecularly imprinted polymers (MIPs): sensing, an explosive new opportunity? Org Biomol Chem. 2007;5(20):3233–44.

[68]   Xu S, Lu H, Zheng X, Chen L. Stimuli-responsive molecularly imprinted polymers: versatile functional materials. J Mater Chem C. 2013;1(29):4406–22.

[69]   Spychalska K, Zajac D, Baluta S, Halicka K, Cabaj J. Functional polymers structures for (Bio) sensing application-a review. Polymers. 2020;12(5):1154.

[70]   Guembe-García M, Peredo-Guzmán PD, Santaolalla-García V, Moradillo-Renuncio N, Ibeas S, Mendía A, et al. Why is the sensory response of organic probes within a polymer film different in solution and in the solid-state? Evidence and application to the detection of amino acids in human chronic wounds. Polymers. 2020;12(6):1249.

[71]   Tyler McQuade D, Pullen AE, Swager TM. Conjugated polymer-based chemical sensors. Chem Rev. 2000;100(7):2537–74.

[72]   Thomas SW, Joly GD, Swager TM. Chemical sensors based on amplifying fluorescent conjugated polymers. Chem Rev. 2007;107(4):1339–86.

[73]   Janata J, Josowicz M. Conducting polymers in electronic chemical sensors. Nat Mater. 2003; 2(1):19–24.

[74]   Gerard M, Chaubey A, Malhotra BD. Application of conducting polymers to biosensors. Biosens Bioelectron. 2002;17(5):345–59.

[75]   Wang C, Jiang T, Zhao K, Deng A, Li J. A novel electrochemiluminescent immunoassay for diclofenac using conductive polymer functionalized graphene oxide as labels and gold nanorods as signal enhancers. Talanta. 2019;193:184–91.

[76]   Jiang X, Wang H, Yuan R, Chai Y. Functional Three-dimensional porous conductive polymer hydrogels for sensitive electrochemiluminescence in situ detection of H_2O_2 released from live cells. Anal Chem. 2018;90(14):8462–9.

[77]   Wasielewski MR. Sensors using molecular recognition in luminescent, conductive polymers [Internet]. US Dep Energy Off Sci Tech Inf. 1999; https://digital.library.unt.edu/ark:/67531/metadc780520/m2/1/high_res_d/828084.pdf.

[78]   Bhadra S, Khastgir D, Singha NK, Lee JH. Progress in preparation, processing and applications of polyaniline. Prog Polym Sci. 2009;34(8):783–810.

[79]   Bahraeian S, Abron K, Pourjafarian F, Majid RA. Study on synthesis of polypyrrole via chemical polymerization method. Adv Mater Res. 2013;795:707–10.

[80]   Brédas JL, Silbey R, Boudreaux DS, Chance RR. Chain-length dependence of electronic and electrochemical properties of conjugated systems: polyacetylene, polyphenylene, polythiophene, and polypyrrole. J Am Chem Soc. 1983;105(22):6555–9.

[81]   Groenendaal L, Jonas F, Freitag D, Pielartzik H, Reynolds JR. Poly(3,4-ethylenedioxythiophene) and its derivatives: past, present, and future. Adv Mater. 2000; 12(7):481–94.

[82]   Zampetti E, Pantalei S, Muzyczuk A, Bearzotti A, De Cesare F, Spinella C, et al. A high sensitive NO_2 gas sensor based on PEDOT-PSS/TiO_2 nanofibres. Sens Actuators B. 2013; 176(2):390–8.

[83] Lange U, Roznyatovskaya N V., Mirsky VM. Conducting polymers in chemical sensors and arrays. Anal Chim Acta. 2008;614(1):1–26.

[84] Borole DD, Kapadi UR, Mahulikar PP, Hundiwale DG. Conducting polymers: an emerging field of biosensors. Des Monomers Polym. 2006;9(1):1–11.

[85] Vallejos S, Hernando E, Trigo M, García FC, García-Valverde M, Iturbe D, et al. Polymeric chemosensor for the detection and quantification of chloride in human sweat. Application to the diagnosis of cystic fibrosis. J Mater Chem B. 2018;6(22):3735–41.

[86] García JM, García FC, Trigo-lópez M, Vallejos S. Tejidos inteligentes Aplicación en detección visual de especies químicas de interés en seguridad laboral y civil, medioambiental e industrial. Revista de Plásticos Modernos. 2016;111(708):6–12.

[87] Vallejos S, Muñoz A, Ibeas S, Serna F, García FC, García JM. Solid sensory polymer substrates for the quantification of iron in blood, wine and water by a scalable RGB technique. J Mater Chem A. 2013;1(48):15435–41.

[88] Vallejos S, Estévez P, García FC, Serna F, De La Peña JL, García JM. Putting to work organic sensing molecules in aqueous media: fluorene derivative-containing polymers as sensory materials for the colorimetric sensing of cyanide in water. Chem Commun. 2010;46 (42):7951–3.

[89] Vallejos S, Muñoz A, García FC, Colleoni R, Biesuz R, Alberti G, et al. Colorimetric detection, quantification and extraction of Fe(III) in water by acrylic polymers with pendant Kojic acid motifs. Sens Actuators B. 2016;233:120–6.

[90] Bustamante SE, Vallejos S, Pascual-Portal BS, Muñoz A, Mendia A, Rivas BL, et al. Polymer films containing chemically anchored diazonium salts with long-term stability as colorimetric sensors. J Hazard Mater. 2019 Mar 5;365:725–32.

[91] Vallejos S, Moreno D, Ibeas S, Muñoz A, García FC, García JM. Polymeric chemosensor for the colorimetric determination of the total polyphenol index (TPI) in wines. Food Control. 2019;106:106684.

[92] González-Ceballos L, Cavia M del M, Fernández-Muiño MA, Osés SM, Sancho MT, Ibeas S, et al. A simple one-pot determination of both total phenolic content and antioxidant activity of honey by polymer chemosensors. Food Chem. 2021;342:12830.

[93] Bamberger E, Padova R, Ormerod E. Über nitro- und amino-formazyl. Justus Liebigs Ann Chem. 1926;446(1):260–307.

[94] Pascual BS, Vallejos S, Reglero Ruiz JA, Bertolín JC, Represa C, García FC, et al. Easy and inexpensive method for the visual and electronic detection of oxidants in air by using vinylic films with embedded aniline. J Hazard Mater. 2019;364:238–43.

[95] Lottenberg R, Christensen U, Jackson CM, Coleman PL. Assay of coagulation proteases using peptide chromogenic and fluorogenic substrates. Methods Enzymol. 1981;80(C):341–61.

[96] Duncan R, Cable H, Lloyd J, Rejmanová P, Kopeček J. Polymers containing enzymatically degradable bonds, 7. Design of oligopeptide side-chains in poly[N-(2-hydroxypropyl) methacrylamide] copolymers to promote efficient degradation by lysosomal enzymes. Makromol Chem. 1983;184(10):1997–2008.

[97] Kopeček J, Rejmanová P, Chytrý V. Polymers containing enzymatically degradable bonds, 1. Chymotrypsin catalyzed hydrolysis of p-nitroanilides of phenylalanine and tyrosine attached to side-chains of copolymers of N-(2-hydroxypropyl)methacrylamide. Makrom Chem. 1981;182(3):799–809.

[98] Subr V, Strohalm J, Ulbrich K, Duncan R, Hume IC. Polymers containing enzymatically degradable bonds, XII. Effect of spacer structure on the rate of release of daunomycin and adriamycin from poly [N-(2-hydroxypropyl)-methacrylamide] copolymer drag carriers in vitro and antitumour activity measured in viv. J Control Release. 1992;18(2):123–32.

[99] Rejmanová P, Kopeček J, Pohl J, Baudyš M, Kostka V. Polymers containing enzymatically degradable bonds, 8. Degradation of oligopeptide sequences in *N*-(2-hydroxypropyl) methacrylamide copolymers by bovine spleen cathepsin B. Makromol Chem. 1983; 184(10):2009–20.

[100] Kopeček, J. Reactive Copolymers of *N*-(2-hydroxypropyl)Methacrylamide with *N*-methacryloylated derivatives of L-Leucine and L-Phenylalanine, 1. Preparation, characterization, and reactions with diamines. Macromol. Chem. Phys. 1977;178(8): 2169–2183.

[101] Drobník J, Kopeček J, Labský J, Rejmanová P, Exner J, Saudek V, et al. Enzymatic cleavage of side chains of synthetic water-soluble polymers. Makromol Chem. 1976;177(10):2833–48.

[102] Selvin PR. The renaissance of fluorescence resonance energy transfer. Nat Struct Biol. 2000;7(9):730–4.

[103] Sapsford KE, Berti L, Medintz IL. Materials for fluorescence resonance energy transfer analysis: Beyond traditional donor-acceptor combinations. Angew Chem Int Ed. 2006; 45(28):4562–89.

[104] Komatsu N, Aoki K, Yamada M, Yukinaga H, Fujita Y, Kamioka Y, et al. Development of an optimized backbone of FRET biosensors for kinases and GTPases. Mol Biol Cell. 2011; 22(23):4647–56.

[105] Xiao Y, Tan X, Li Z, Zhang K. Self-immolative polymers in biomedicine. J Mater Chem B. 2020;8(31):6697–709.

[106] Sagi A, Weinstain R, Karton N, Shabat D. Self-immolative polymers. J Am Chem Soc. 2008;130(16):5434–5.

[107] Wang TT, Lio C kit, Huang H, Wang RY, Zhou H, Luo P, et al. A feasible image-based colorimetric assay using a smartphone RGB camera for point-of-care monitoring of diabetes. Talanta. 2020;206:120211.

[108] Vallejos S, Guembe García M, García Pérez JM, Represa Pérez C, García García FC. Application for smartphones "Colorimetric Titration" on Google Play Store [Internet]. 2021 [cited 2021 Aug 7]. Available from: https://play.google.com/store/apps/details?id=es.inforapps.chame leon&gl=ES

[109] Vallejos S, Guembe García M, García Pérez JM, Represa Pérez C, García García FC. Application for smartphones "Colorimetric Titration" on the App Store [Internet]. 2021 [cited 2021 Aug 7]. Available from: https://apps.apple.com/si/app/colorimetric-titration/id1533793244.

[110] Frackowiak D. The Jablonski diagram. J Photochem Photobiol B Biol. 1988;2(3):399.

[111] Calvo-Gredilla P, García-Calvo J, Cuevas J V., Torroba T, Pablos JL, García FC, et al. Solvent-free off–on detection of the improvised explosive triacetone triperoxide (TATP) with fluorogenic materials. Chem Eur J. 2017;23(56):13973–9.

[112] Guembe-García M, Vallejos S, Carreira-barral I, Ibeas S, García FC, Santaolalla-García V, et al. Zn(II) detection in biological samples with a smart sensory polymer. React Funct Polym. 2020;154:104685.

[113] Rahman MA, Kumar P, Park DS, Shim YB. Electrochemical sensors based on organic conjugated polymers. Sensors. 2008;8(1):118–41.

[114] Blanco-López MC, Lobo-Castañón MJ, Miranda-Ordieres AJ, Tuñón-Blanco P. Electrochemical sensors based on molecularly imprinted polymers. Trends Anal Chem. 2004;23(1):36–48.

[115] Bidan G. Electroconducting conjugated polymers: new sensitive matrices to build up chemical or electrochemical sensors. A review. Sens Actuators B. 1992;6(1–3):45–56.

[116] Hangarter CM, Chartuprayoon N, Hernández SC, Choa Y, Myung N V. Hybridized conducting polymer chemiresistive nano-sensors. Nano Today. 2013;8(1):39–55.

[117] Macdiarmid AG, Mammone RJ, Kaner RB, Porter SJ, Pethig R, Heeger AJ, et al. The concept of 'doping' of conducting polymers: the role of reduction potentials. Philos Trans R Soc London Ser A, Math Phys Sci. 1985;314(1528):3–15.

[118] Wong YC, Ang BC, Haseeb ASMA, Baharuddin AA, Wong YH. Review—conducting polymers as chemiresistive gas sensing materials: a review. J Electrochem Soc. 2020;167(3):037503.

[119] Bouvet M, Mateos M, Meunier-Prest R, Suisse J-M. Conducting polymers for ammonia sensing: electrodeposition, hybrid materials and heterojunctions. Proceedings. 2017; 1(4):480.

[120] Guernion N, Ewen RJ, Pihlainen K, Ratcliffe NM, Teare GC. The fabrication and characterisation of a highly sensitive polypyrrole sensor and its electrical responses to amines of differing basicity at high humidities. Synth Met. 2002;126(2–3):301–10.

[121] Cho JH, Yu JB, Kim JS, Sohn SO, Lee DD, Huh JS. Sensing behaviors of polypyrrole sensor under humidity condition. Sens Actuators B. 2005;108(1–2):389–92.

[122] Jeon SS, An HH, Yoon CS, Im SS. Synthesis of ultra-thin polypyrrole nanosheets for chemical sensor applications. Polymer. 2011;52(3):652–7.

[123] Sengupta PP, Barik S, Adhikari B. Polyaniline as a gas-sensor material. Mater Manuf Process. 2006;21(3):263–70.

[124] Sengupta PP, Adhikari B. Influence of polymerization condition on the electrical conductivity and gas sensing properties of polyaniline. Mater Sci Eng A. 2007;459(1–2):278–85.

[125] Tiwari A, Kumar R, Prabaharan M, Pandey RR, Kumari P, Chaturvedi A, et al. Nanofibrous polyaniline thin film prepared by plasma-induced polymerization technique for detection of NO_2 gas. Polym Adv Technol. 2010;21(9):615–20.

[126] Rella R, Siciliano P, Quaranta F, Primo T, Valli L, Schenetti L, et al. Gas sensing measurements and analysis of the optical properties of poly[3-(butylthio)thiophene] Langmuir-Blodgett films. Sens Actuators B. 2000;68(1):203–9.

[127] Lu HH, Lin CY, Hsiao TC, Fang YY, Ho KC, Yang D, et al. Electrical properties of single and multiple poly(3,4-ethylenedioxythiophene) nanowires for sensing nitric oxide gas. Anal Chim Acta. 2009;640(1–2):68–74.

[128] Wang X, Lv Y, Hou X. A potential visual fluorescence probe for ultratrace arsenic (III) detection by using glutathione-capped CdTe quantum dots. Talanta. 2011;84(2):382–6.

[129] Zhang X, Yin J, Yoon J. Recent advances in development of chiral fluorescent and colorimetric sensors. Chem Rev. 2014;114(9):4918–59.

[130] Martín-Yerga D, González-García MB, Costa-García A. Electrochemical determination of mercury: a review. Talanta. 2013;116:1091–104.

[131] Wang H, Xu C, Yuan B. Polymer-based electrochemical sensing platform for heavy metal ions detection - a critical review. Int J Electrochem Sci. 2019;14(9):8760–71.

[132] Joseph A, Subramanian S, Ramamurthy PC, Sampath S, Kumar RV, Schwandt C. Iminodiacetic acid functionalized polypyrrole modified electrode as Pb(II) sensor: synthesis and DPASV studies. Electrochim Acta. 2014;137:557–63.

[133] Zuo Y, Xu J, Zhu X, Duan X, Lu L, Gao Y, et al. Poly(3,4-ethylenedioxythiophene) nanorods/ graphene oxide nanocomposite as a new electrode material for the selective electrochemical detection of mercury (II). Synth Met. 2016;220:14–9.

[134] Deshmukh MA, Celiesiute R, Ramanaviciene A, Shirsat MD, Ramanavicius A. EDTA_PANI/ SWCNTs nanocomposite modified electrode for electrochemical determination of copper (II), lead (II) and mercury (II) ions. Electrochim Acta. 2018;259:930–8.

[135] Alizadeh T, Amjadi S. Preparation of nano-sized Pb^{2+} imprinted polymer and its application as the chemical interface of an electrochemical sensor for toxic lead determination in different real samples. J Hazard Mater. 2011;190(1–3):451–9.

[136] Topcu C, Lacin G, Yilmaz V, Coldur F, Caglar B, Cubuk O, et al. Electrochemical determination of copper(II) in water samples using a novel ion-selective electrode based on a graphite oxide–imprinted polymer composite. Anal Lett. 2018;51(12):1890–910.

[137] Zhang YX, Zhao PY, Yu LP. Highly-sensitive and selective colorimetric sensor for amino acids chiral recognition based on molecularly imprinted photonic polymers. Sens Actuators B. 2013;181:850–7.

[138] Van Grinsven B, Eersels K, Akkermans O, Ellermann S, Kordek A, Peeters M, et al. Label-free detection of *Escherichia coli* based on thermal transport through surface imprinted polymers. ACS Sens. 2016;1(9):1140–7.

[139] Zhou J, Yao D, Qian Z, Hou S, Li L, Jenkins ATA, et al. Bacteria-responsive intelligent wound dressing: Simultaneous In situ detection and inhibition of bacterial infection for accelerated wound healing. Biomaterials. 2018;161:11–23.

[140] Feng S, Hu Y, Ma L, Lu X. Development of molecularly imprinted polymers-surface-enhanced Raman spectroscopy/colorimetric dual sensor for determination of chlorpyrifos in apple juice. Sens Actuators B. 2017 Mar 31;241:750–7.

[141] Liang R, Zhang R, Qin W. Potentiometric sensor based on molecularly imprinted polymer for determination of melamine in milk. Sens Actuators B. 2009;141(2):544–50.

[142] Li B, Zhang Z, Qi J, Zhou N, Qin S, Choo J, et al. Quantum dot-based molecularly imprinted polymers on three-dimensional origami paper microfluidic chip for fluorescence detection of phycocyanin. ACS Sens. 2017;2(2):243–50.

[143] Isaad J, Salaün F. Functionalized poly (vinyl alcohol) polymer as chemodosimeter material for the colorimetric sensing of cyanide in pure water. Sens Actuators B. 2011;157(1):26–33.

[144] Kong Q, Wang Y, Zhang L, Ge S, Yu J. A novel microfluidic paper-based colorimetric sensor based on molecularly imprinted polymer membranes for highly selective and sensitive detection of bisphenol A. Sens Actuators B. 2017;243:130–6.

[145] Waghuley SA, Yenorkar SM, Yawale SS, Yawale SP. Application of chemically synthesized conducting polymer-polypyrrole as a carbon dioxide gas sensor. Sens Actuators B. 2008;128(2):366–73.

[146] Yan XB, Han ZJ, Yang Y, Tay BK. NO$_2$ gas sensing with polyaniline nanofibers synthesized by a facile aqueous/organic interfacial polymerization. Sens Actuators B. 2007;123(1):107–13.

[147] Ram MK, Yavuz Ö, Lahsangah V, Aldissi M. CO gas sensing from ultrathin nano-composite conducting polymer film. Sens Actuators B. 2005;106(2):750–7.

[148] Tanwar AS, Hussain S, Malik AH, Afroz MA, Iyer PK. Inner filter effect based selective detection of nitroexplosive-picric acid in aqueous solution and solid support using conjugated polymer. ACS Sens. 2016;1(8):1070–7.

[149] Pablos JL, Trigo-López M, Serna F, García FC, García JM. Water-soluble polymers, solid polymer membranes, and coated fibres as smart sensory materials for the naked eye detection and quantification of TNT in aqueous media. Chem Commun. 2014;50(19):2484–7.

[150] Pablos JL, Trigo-López M, Serna F, García FC, García JM. Solid polymer substrates and smart fibres for the selective visual detection of TNT both in vapour and in aqueous media. RSC Adv. 2014;4(49):25562–8.

[151] Pablos JL, Estévez P, Muñoz A, Ibeas S, Serna F, García FC, et al. Polymer chemosensors as solid films and coated fibres for extreme acidity colorimetric sensing. J Mater Chem A. 2015;3(6):2833–43.

[152] Trigo-López M, Pablos JL, Muñoz A, Ibeas S, Serna F, García FC, et al. Aromatic polyamides and acrylic polymers as solid sensory materials and smart coated fibres for high acidity colorimetric sensing. Polym Chem. 2015;6(16):3110–20.

[153] Park M. orientation control of the molecular recognition layer for improved sensitivity: a review. Biochip J. 2019;13(1):82–94.

[154] Peltomaa R, Glahn-Martínez B, Benito-Peña E, Moreno-Bondi MC. Optical biosensors for label-free detection of small molecules. Sensors. 2018;18(12):4126.

[155] Husain Q. Nanocarriers immobilized proteases and their industrial applications: an overview. J Nanosci Nanotechnol. 2018;18(1):486–99.

[156] Hasan A, Pandey LM. Review: polymers, surface-modified polymers, and self assembled monolayers as surface-modifying agents for biomaterials. Polym Plast Technol Eng. 2015; 54(13):1358–78.

[157] Vashist SK, Lam E, Hrapovic S, Male KB, Luong JHT. Immobilization of antibodies and enzymes on 3-aminopropyltriethoxysilane-functionalized bioanalytical platforms for biosensors and diagnostics. Chem Rev. 2014;114(21):11083–130.

[158] Takai M. Highly sensitive and rapid biosensing on a three-dimensional polymer platform. Polym J. 2018;50(9):847–55.

[159] Li Z, Chen GY. Current conjugation methods for immunosensors. Nanomaterials. 2018;8(5): 1–11.

[160] Sharma SK, Leblanc RM. Biosensors based on β-galactosidase enzyme: recent advances and perspectives. Anal Biochem. 2017;535:1–11.

[161] Liu Y, Yu J. Oriented immobilization of proteins on solid supports for use in biosensors and biochips: a review. Microchim Acta. 2016;183(1):1–19.

[162] Welch NG, Scoble JA, Muir BW, Pigram PJ. Orientation and characterization of immobilized antibodies for improved immunoassays. Biointerphases. 2017;12(2):02D301.

[163] Yu Q, Wang Q, Li B, Lin Q, Duan Y. Technological development of antibody immobilization for optical immunoassays: progress and prospects. Crit Rev Anal Chem. 2015;45(1):62–75.

[164] Wang C, Feng B. Research progress on site-oriented and three-dimensional immobilization of protein. Mol Biol. 2015;49(1):1–20.

[165] Faccio G. From protein features to sensing surfaces. Sensors. 2018;18(4):1204.

[166] Trilling AK, Hesselink T, Houwelingen A van, Cordewener JHG, Jongsma MA, Schoffelen S, et al. Orientation of llama antibodies strongly increases sensitivity of biosensors. Biosens Bioelectron. 2014;60:130–6.

[167] Bereli N, Ertürk G, Tümer MA, Say R, Denizli A. Oriented immobilized anti-hIgG via Fc fragment-imprinted PHEMA cryogel for IgG purification. Biomed Chromatogr. 2013;27(5): 599–607.

[168] Orlov A V., Bragina VA, Nikitin MP, Nikitin PI. Rapid dry-reagent immunomagnetic biosensing platform based on volumetric detection of nanoparticles on 3D structures. Biosens Bioelectron. 2016;79:423–9.

[169] Adak AK, Li BY, Huang L De, Lin TW, Chang TC, Hwang KC, et al. Fabrication of antibody microarrays by light-induced covalent and oriented immobilization. ACS Appl Mater Interfaces. 2014;6(13):10452–60.

[170] Lebec V, Boujday S, Poleunis C, Pradier CM, Delcorte A. Time-of-flight secondary ion mass spectrometry investigation of the orientation of adsorbed antibodies on SAMs correlated to biorecognition tests. J Phys Chem C. 2014;118(4):2085–92.

[171] Ulman A. Formation and structure of self-assembled monolayers. Chem Rev. 1996; 96(4):1533–54.

[172] Borges J, Mano JF. Molecular interactions driving the layer-by-layer assembly of multilayers. Chem Rev. 2014;114(18):8883–942.

[173] Zasadzinski JA, Viswanathan R, Madsen L, Garnaes J, Schwartz DK. Langmuir-Blodgett Films. Science. 1994;263(5154):1726–33.

[174] Reitzel N, Greve DR, Kjaer K, Howes PB, Jayaraman M, Savoy S, et al. Self-assembly of conjugated polymers at the air/water interface. Structure and properties of Langmuir and

Langmuir-Blodgett films of amphiphilic regioregular polythiophenes. J Am Chem Soc. 2000;122(24):5788–800.

[175] L Krówczyński. The development of pharmaceutical technology (chronological tabulated facts). Pharmazie. 1985;40(5):346–9.

[176] Chien YW. Novel drug delivery systems. 2nd ed. Dekker M, editor. New York; 1992.

[177] Yun YH, Lee BK, Park K. Controlled drug delivery: historical perspective for the next generation. J Control Release. 2015;219:2–7.

[178] Dokoumetzidis A, Macheras P. A century of dissolution research: from Noyes and Whitney to the biopharmaceutics classification system. Int J Pharm. 2006;321(1–2):1–11.

[179] Wilding IR. Site-specific drug delivery in the gastrointestinal tract. Crit Rev Ther Drug Carr Syst. 2000;17(6):76.

[180] Rathbone MJ, Hadgraft J, Roberts MS. Modified-release drug delivery technology. CRC Press; 2002.

[181] Meyer DE, Shin BC, Kong GA, Dewhirst MW, Chilkoti A. Drug targeting using thermally responsive polymers and local hyperthermia. J Control Release. 2001;74(1–3):213–24.

[182] Kim S, Kim J-H, Jeon O, Kwon IC, Park K. Engineered polymers for advanced drug delivery. Eur J Pharm Biopharm. 2009;71(3):420–30.

[183] Kost, J. Intelligent Drug Delivery Systems, in: Mathiowitz, E, editor. The encyclopedia of controlled drug delivery. John Wiley & Sons; 2000. pp 445-459.

[184] Koo OM, Varia SA. Case studies with new excipients: development, implementation and regulatory approval. Ther Deliv. 2011;2(7):949–56.

[185] Alvarez-Lorenzo C, Concheiro A. Intelligent drug delivery systems: polymeric micelles and hydrogels. Mini Rev Med Chem. 2008;8(11):1065–74.

[186] Roy I, Gupta MN. Smart polymeric materials. Chem Biol. 2003;10(12):1161–71.

[187] Voit B, Appelhans D. Glycopolymers of various architectures-more than mimicking nature. Macromol Chem Phys. 2010;211(7):727–35.

[188] Alvarez-Lorenzo C, Concheiro A. From drug dosage forms to intelligent drug-delivery systems: a change of paradigm. In: Schneider H-J, Shahinpoor M, editors. Smart materials for drug delivery: Volume 1. Royal Society of Chemistry; 2013. p. 1–32.

[189] Youan B-BC. Chronopharmaceutics: gimmick or clinically relevant approach to drug delivery? J Control Release. 2004;98(3):337–53.

[190] Lévi F, Okyar A. Circadian clocks and drug delivery systems: impact and opportunities in chronotherapeutics. Expert Opin Drug Deliv. 2011;8(12):1535–41.

[191] Lakkadwala S, Nguyen S, Nesamony J, Narang AS, Boddu SHS. Smart polymers in drug delivery. In: Narang AS, Boddu SHS, editors. Excipient applications in formulation design and drug delivery. Cham: Springer International Publishing; 2015. p. 169–99.

[192] Mccarthy W. Polymeric drug delivery techniques. Aldrich Mater Sci. 2016;3–12.

[193] Semwal R, Semwal RB, Semwal DK. Drug delivery systems: selection criteria and use. In: Zohuri G, Ahmadjoo S, Rabiee A, Shamakhi MA, editors. Encyclopedia of biomedical polymers and polymeric biomaterials. Taylor & Francis; 2015. p. 2938–49.

[194] Mishra RK, Tiwari SK, Mohapatra S, Thomas S. Efficient nanocarriers for drug-delivery systems. In: Mohapatra SS, Ranjan S, Dasgupta N, Mishra RK, Thomas S, editors. Nanocarriers for drug delivery. Elsevier Inc.; 2019. p. 1–41.

[195] Yadav HKS, Almokdad AA, Shaluf SIM, Debe MS. Polymer-based nanomaterials for drug-delivery carriers. In: Mohapatra SS, Ranjan S, Dasgupta N, Mishra RK, Sabu Thomas, editors. Nanocarriers for drug delivery. Elsevier; 2019. p. 531–56.

[196] Priya James H, John R, Alex A, Anoop KR. Smart polymers for the controlled delivery of drugs – a concise overview. Acta Pharm Sin B. 2014;4(2):120–7.

[197] Wells CM, Harris M, Choi L, Murali VP, Guerra FD, Jennings JA. Stimuli-responsive drug release from smart polymers. J Funct Biomater. 2019;10(3):34.

[198] Aghabegi Moghanjoughi A, Khoshnevis D, Zarrabi A. A concise review on smart polymers for controlled drug release. Drug Deliv Transl Res. 2016;6(3):333–40.

[199] Aguilar MR, San Román J. Introduction to smart polymers and their applications. In: Aguilar MR, San Román J, editors. Smart polymers and their applications. 2nd ed. Woodhead Publishing; 2014. Ch. 1, pp. 1–11.

[200] Alsehli M. Polymeric nanocarriers as stimuli-responsive systems for targeted tumor (cancer) therapy: recent advances in drug delivery. Saudi Pharm J. 2020;28(3):255–65.

[201] Van Hove A, Cui Z, Benoit D. Stimuli-responsive polymer delivery systems. In: Bader RA, Putnam DA, editors. Engineering polymer systems for improved drug delivery. Hoboken, NJ, USA: John Wiley & Sons, Inc.; 2013. p. 377–427.

[202] Sponchioni M, Capasso Palmiero U, Moscatelli D. Thermo-responsive polymers: applications of smart materials in drug delivery and tissue engineering. Mater Sci Eng C. 2019;102: 589–605.

[203] Chaterji S, Kwon IK, Park K. Smart polymeric gels: redefining the limits of biomedical devices. Prog Polym Sci. 2007;32(8–9):1083–122.

[204] Guo H, Mussault C, Marcellan A, Hourdet D, Sanson N. Hydrogels with Dual Thermoresponsive Mechanical Performance. Macromol Rapid Commun. 2017;38 (17):1700287.

[205] Panayiotou M, Pöhner C, Vandevyver C, Wandrey C, Hilbrig F, Freitag R. Synthesis and characterisation of thermo-responsive poly(N,N'-diethylacrylamide) microgels. React Funct Polym. 2007;67(9):807–19.

[206] Kono K, Ozawa T, Yoshida T, Ozaki F, Ishizaka Y, Maruyama K, et al. Highly temperature-sensitive liposomes based on a thermosensitive block copolymer for tumor-specific chemotherapy. Biomaterials. 2010;31(27):7096–105.

[207] Dreher M. Evaluation of an elastin-like polypeptide–doxorubicin conjugate for cancer therapy. J Control Release. 2003;91(1–2):31–43.

[208] Luxenhofer R, Schulz A, Roques C, Li S, Bronich TK, Batrakova E V., et al. Doubly amphiphilic poly(2-oxazoline)s as high-capacity delivery systems for hydrophobic drugs. Biomaterials. 2010;31(18):4972–9.

[209] Cheng C-C, Liang M-C, Liao Z-S, Huang J-J, Lee D-J. Self-assembled supramolecular nanogels as a safe and effective drug delivery vector for cancer therapy. Macromol Biosci. 2017; 17(5):1600370.

[210] Wang Q, Li S, Wang Z, Liu H, Li C. Preparation and characterization of a positive thermoresponsive hydrogel for drug loading and release. J Appl Polym Sci. 2009; 111(3):1417–25.

[211] Li W, Huang L, Ying X, Jian Y, Hong Y, Hu F, et al. Antitumor drug delivery modulated by a polymeric micelle with an upper critical solution temperature. Angew Chem Int Ed. 2015; 54(10):3126–31.

[212] Hei M, Wang J, Wang K, Zhu W, Ma PX. Dually responsive mesoporous silica nanoparticles regulated by upper critical solution temperature polymers for intracellular drug delivery. J Mater Chem B. 2017;5(48):9497–501.

[213] Li S, Hu K, Cao W, Sun Y, Sheng W, Li F, et al. pH-responsive biocompatible fluorescent polymer nanoparticles based on phenylboronic acid for intracellular imaging and drug delivery. Nanoscale. 2014;6(22):13701–9.

[214] Qu J, Zhao X, Ma PX, Guo B. pH-responsive self-healing injectable hydrogel based on N-carboxyethyl chitosan for hepatocellular carcinoma therapy. Acta Biomater. 2017;58:168–80.

[215] Chung M-F, Chia W-T, Liu H-Y, Hsiao C-W, Hsiao H-C, Yang C-M, et al. Inflammation-induced drug release by using a pH-responsive gas-generating hollow-microsphere system for the treatment of osteomyelitis. Adv Healthc Mater. 2014;3(11):1854–61.

[216] Kulkarni R V., Boppana R, Krishna Mohan G, Mutalik S, Kalyane N V. pH-responsive interpenetrating network hydrogel beads of poly(acrylamide)-g-carrageenan and sodium alginate for intestinal targeted drug delivery: synthesis, in vitro and in vivo evaluation. J Colloid Interface Sci. 2012;367(1):509–17.

[217] Liao J, Zheng H, Fei Z, Lu B, Zheng H, Li D, et al. Tumor-targeting and pH-responsive nanoparticles from hyaluronic acid for the enhanced delivery of doxorubicin. Int J Biol Macromol. 2018;113:737–47.

[218] Go Y-M, Jones DP. Redox compartmentalization in eukaryotic cells. Biochim Biophys Acta - Gen Subj. 2008;1780(11):1273–90.

[219] Quinn JF, Whittaker MR, Davis TP. Glutathione responsive polymers and their application in drug delivery systems. Polym Chem. 2017;8(1):97–126.

[220] Guo X, Cheng Y, Zhao X, Luo Y, Chen J, Yuan WE. Advances in redox-responsive drug delivery systems of tumor microenvironment. J Nanobiotechnology. 2018;16(1):1–10.

[221] Velluto D, Thomas SN, Simeoni E, Swartz MA, Hubbell JA. PEG-b-PPS-b-PEI micelles and PEG-b-PPS/PEG-b-PPS-b-PEI mixed micelles as non-viral vectors for plasmid DNA: tumor immunotoxicity in B16F10 melanoma. Biomaterials. 2011;32(36):9839–47.

[222] Song N, Liu W, Tu Q, Liu R, Zhang Y, Wang J. Preparation and in vitro properties of redox-responsive polymeric nanoparticles for paclitaxel delivery. Colloids Surf B. 2011; 87(2):454–63.

[223] Zhuang Y, Deng H, Su Y, He L, Wang R, Tong G, et al. Aptamer-functionalized and backbone redox-responsive hyperbranched polymer for targeted drug delivery in cancer therapy. Biomacromolecules. 2016;17(6):2050–62.

[224] Maiti C, Parida S, Kayal S, Maiti S, Mandal M, Dhara D. Redox-responsive core-cross-linked block copolymer micelles for overcoming multidrug resistance in cancer cells. ACS Appl Mater Interfaces. 2018;10(6):5318–30.

[225] Park W, Bae B, Na K. A highly tumor-specific light-triggerable drug carrier responds to hypoxic tumor conditions for effective tumor treatment. Biomaterials. 2016;77:227–34.

[226] Higuchi A, Ling Q-D, Kumar SS, Chang Y, Kao T-C, Munusamy MA, et al. External stimulus-responsive biomaterials designed for the culture and differentiation of ES, iPS, and adult stem cells. Prog Polym Sci. 2014;39(9):1585–613.

[227] Poelma SO, Oh SS, Helmy S, Knight AS, Burnett GL, Soh HT, et al. Controlled drug release to cancer cells from modular one-photon visible light-responsive micellar system. Chem Commun. 2016;52(69):10525–8.

[228] Jin Q, Cai T, Han H, Wang H, Wang Y, Ji J. Light and pH dual-degradable triblock copolymer micelles for controlled intracellular drug release. Macromol Rapid Commun. 2014; 35(15):1372–8.

[229] Alvarez-Lorenzo C, Bromberg L, Concheiro A. Light-sensitive intelligent drug delivery systems. Photochem Photobiol. 2009;85(4):848–60.

[230] Sortino S. Photoactivated nanomaterials for biomedical release applications. J Mater Chem. 2012;22(2):301–18.

[231] Xiong X, del Campo A, Cui J. Photoresponsive polymers. In: Aguilar MR, San Román J, editors. Smart Polymers and their Applications. Elsevier; 2019. Ch. 4, pp. 87–153.

[232] Sun T, Li P, Oh JK. Dual location dual reduction/photoresponsive block copolymer micelles: disassembly and synergistic release. Macromol Rapid Commun. 2015;36(19):1742–8.

[233] Rastogi SK, Anderson HE, Lamas J, Barret S, Cantu T, Zauscher S, et al. Enhanced release of molecules upon ultraviolet (UV) light irradiation from photoresponsive hydrogels prepared

from bifunctional azobenzene and four-arm Poly(ethylene glycol). ACS Appl Mater Interfaces. 2018;10(36):30071–80.

[234] Xiang J, Tong X, Shi F, Yan Q, Yu B, Zhao Y. Near-infrared light-triggered drug release from UV-responsive diblock copolymer-coated upconversion nanoparticles with high monodispersity. J Mater Chem B. 2018;6(21):3531–40.

[235] Son S, Shin E, Kim B-S. Light-responsive micelles of spiropyran initiated hyperbranched polyglycerol for smart drug delivery. Biomacromolecules. 2014;15(2):628–34.

[236] Pearson S, Vitucci D, Khine YY, Dag A, Lu H, Save M, et al. Light-responsive azobenzene-based glycopolymer micelles for targeted drug delivery to melanoma cells. Eur Polym J. 2015;69:616–27.

[237] Xing Q, Li N, Chen D, Sha W, Jiao Y, Qi X, et al. Light-responsive amphiphilic copolymer coated nanoparticles as nanocarriers and real-time monitors for controlled drug release. J Mater Chem B. 2014;2(9):1182.

[238] Anal A. Stimuli-induced pulsatile or triggered release delivery systems for bioactive compounds. Recent Pat Endocr Metab Immune Drug Discov. 2007;1(1):83–90.

[239] Murdan S. Electro-responsive drug delivery from hydrogels. J Control Release. 2003; 92(1–2):1–17.

[240] Ramanathan S, Block LH. The use of chitosan gels as matrices for electrically-modulated drug delivery. J Control Release. 2001;70(1–2):109–23.

[241] KagataniX S, Shinoda T, Konno Y, Fukui M, Ohmura T, Osada Y. Electroresponsive pulsatile depot delivery of insulin from poly(dimethylaminopropylacrylamide) gel in rats. J Pharm Sci. 1997;86(11):1273–7.

[242] Svirskis D, Travas-Sejdic J, Rodgers A, Garg S. Electrochemically controlled drug delivery based on intrinsically conducting polymers. J Control Release. 2010;146(1):6–15.

[243] Balint R, Cassidy NJ, Cartmell SH. Conductive polymers: Towards a smart biomaterial for tissue engineering. Acta Biomater. 2014;10(6):2341–53.

[244] Mongkolkitikul S, Paradee N, Sirivat A. Electrically controlled release of ibuprofen from conductive poly(3-methoxydiphenylamine)/crosslinked pectin hydrogel. Eur J Pharm Sci. 2018;112:20–7.

[245] Lee H, Hong W, Jeon S, Choi Y, Cho Y. Electroactive polypyrrole nanowire arrays: synergistic effect of cancer treatment by on-demand drug release and photothermal therapy. Langmuir. 2015;31(14):4264–9.

[246] Atoufi Z, Zarrintaj P, Motlagh GH, Amiri A, Bagher Z, Kamrava SK. A novel bio electro active alginate-aniline tetramer/ agarose scaffold for tissue engineering: synthesis, characterization, drug release and cell culture study. J Biomater Sci Polym Ed. 2017; 28(15):1617–38.

[247] Krukiewicz K, Zawisza P, Herman AP, Turczyn R, Boncel S, Zak JK. An electrically controlled drug delivery system based on conducting poly(3,4-ethylenedioxypyrrole) matrix. Bioelectrochemistry. 2016;108:13–20.

[248] Agnihotri SA, Kulkarni R V., Mallikarjuna NN, Kulkarni P V., Aminabhavi TM. Electrically modulated transport of diclofenac salts through hydrogels of sodium alginate, Carbopol, and their blend polymers. J Appl Polym Sci. 2005;96(2):301–11.

[249] Arruebo M, Fernández-Pacheco R, Ibarra MR, Santamaría J. Magnetic nanoparticles for drug delivery. Nano Today. 2007;2(3):22–32.

[250] Kumar CSSR, Mohammad F. Magnetic nanomaterials for hyperthermia-based therapy and controlled drug delivery. Adv Drug Deliv Rev. 2011;63(9):789–808.

[251] Liu T-Y, Hu S-H, Liu D-M, Chen S-Y, Chen I-W. Biomedical nanoparticle carriers with combined thermal and magnetic responses. Nano Today. 2009;4(1):52–65.

[252] Caravan P, Ellison JJ, McMurry TJ, Lauffer RB. Gadolinium(III) Chelates as MRI contrast agents: structure, dynamics, and applications. Chem Rev. 1999;99(9):2293–352.

[253] Xu H, Regino CAS, Koyama Y, Hama Y, Gunn AJ, Bernardo M, et al. Preparation and preliminary evaluation of a biotin-targeted, lectin-targeted dendrimer-based probe for dual-modality magnetic resonance and fluorescence imaging. Bioconjug Chem. 2007; 18(5):1474–82.

[254] Pan B, Cui D, Sheng Y, Ozkan C, Gao F, He R, et al. Dendrimer-modified magnetic nanoparticles enhance efficiency of gene delivery system. Cancer Res. 2007;67(17):8156–63.

[255] Riedinger A, Guardia P, Curcio A, Garcia MA, Cingolani R, Manna L, et al. Subnanometer local temperature probing and remotely controlled drug release based on azo-functionalized iron oxide nanoparticles. Nano Lett. 2013;13(6):2399–406.

[256] Duan J, Dong J, Zhang T, Su Z, Ding J, Zhang Y, et al. Polyethyleneimine-functionalized iron oxide nanoparticles for systemic siRNA delivery in experimental arthritis. Nanomedicine. 2014;9(6):789–801.

[257] VandenBerg MA, Webber MJ. Biologically inspired and chemically derived methods for glucose-responsive insulin therapy. Adv Healthc Mater. 2019;8(12):1801466.

[258] James TD, Shinkai S. Artificial receptors as chemosensors for carbohydrates. In: Penades S, editor. Host-guest chemistry mimetic approaches to study carbohydrate recognition. Springer; 2002. p. 159–200.

[259] Tanna S, Joan Taylor M, Sahota TS, Sawicka K. Glucose-responsive UV polymerised dextran-concanavalin A acrylic derivatised mixtures for closed-loop insulin delivery. Biomaterials. 2006;27(8):1586–97.

[260] Ravaine V, Ancla C, Catargi B. Chemically controlled closed-loop insulin delivery. J Control Release. 2008;132(1):2–11.

[261] Gu Z, Dang TT, Ma M, Tang BC, Cheng H, Jiang S, et al. Glucose-responsive microgels integrated with enzyme nanocapsules for closed-loop insulin delivery. ACS Nano. 2013; 7(8):6758–66.

[262] Hu X, Yu J, Qian C, Lu Y, Kahkoska AR, Xie Z, et al. H_2O_2-responsive vesicles integrated with transcutaneous patches for glucose-mediated insulin delivery. ACS Nano. 2017;11(1):613–20.

[263] Yin R, Tong Z, Yang D, Nie J. Glucose and pH dual-responsive concanavalin A based microhydrogels for insulin delivery. Int J Biol Macromol. 2011;49(5):1137–42.

[264] Tanna S, Sahota T, Clark J, Taylor MJ. Covalent coupling of concanavalin A to a Carbopol 934P and 941P carrier in glucose-sensitive gels for delivery of insulin. J Pharm Pharmacol. 2002; 54(11):1461–9.

[265] Kim H, Kang YJ, Kang S, Kim KT. Monosaccharide-responsive release of insulin from polymersomes of polyboroxole block copolymers at neutral pH. J Am Chem Soc. 2012; 134(9):4030–3.

[266] Wang Y, Huang F, Sun Y, Gao M, Chai Z. Development of shell cross-linked nanoparticles based on boronic acid-related reactions for self-regulated insulin delivery. J Biomater Sci Polym Ed. 2017;28(1):93–106.

[267] Liu Y, Ding X, Li J, Luo Z, Hu Y, Liu J, et al. Enzyme responsive drug delivery system based on mesoporous silica nanoparticles for tumor therapy in vivo. Nanotechnology. 2015; 26(14):145102.

[268] Van Tomme SR, Storm G, Hennink WE. In situ gelling hydrogels for pharmaceutical and biomedical applications. Int J Pharm. 2008;355(1–2):1–18.

[269] Zhu L, Wang T, Perche F, Taigind A, Torchilin VP. Enhanced anticancer activity of nanopreparation containing an MMP2-sensitive PEG-drug conjugate and cell-penetrating moiety. Proc Natl Acad Sci. 2013;110(42):17047–52.

[270] Law B, Tung C-H. Proteolysis: a biological process adapted in drug delivery, therapy, and imaging. Bioconjug Chem. 2009;20(9):1683–95.

[271] Bhawania SA, Nisar M, Tariq A, Alotaibi KM, Asaruddin R. Enzyme-responsive polymer composites and their applications. In: Bhawani SA, Khan A, Jawaid M. Smart Polymer Nanocomposites. Woodhead Publishing; 2021. Ch. 7, pp. 169-82.

[272] Wang J, Zhang H, Wang F, Ai X, Huang D, Liu G, et al. Enzyme-responsive polymers for drug delivery and molecular imaging. In: Makhlouf ASH, Abu-Thabit NY, editors. Stimuli responsive polymeric nanocarriers for drug delivery applications: Volume 1: Types and Triggers. Woodhead Publishing; 2018. Ch. 4, pp. 101–19.

[273] Anjum F, Lienemann PS, Metzger S, Biernaskie J, Kallos MS, Ehrbar M. Enzyme responsive GAG-based natural-synthetic hybrid hydrogel for tunable growth factor delivery and stem cell differentiation. Biomaterials. 2016;87:104–17.

[274] Knipe JM, Chen F, Peppas NA. Enzymatic biodegradation of hydrogels for protein delivery targeted to the small intestine. Biomacromolecules. 2015;16(3):962–72.

[275] Rao J, Khan A. Enzyme sensitive synthetic polymer micelles based on the azobenzene motif. J Am Chem Soc. 2013;135(38):14056–9.

[276] Gu X, Qiu M, Sun H, Zhang J, Cheng L, Deng C, et al. Polytyrosine nanoparticles enable ultra-high loading of doxorubicin and rapid enzyme-responsive drug release. Biomater Sci. 2018; 6(6):1526–34.

[277] Miyata T, Uragami T. Biological stimulus-responsive hydrogels. In: Dumitriu S, editor. Polymeric Biomaterials. Taylor & Francis Group; 2002. p. 959–74.

[278] Kokufata E, Zhang Y-Q, Tanaka T. Saccharide-sensitive phase transition of a lectin-loaded gel. Nature. 1991;351:302–4.

[279] Lu Z-R, Kopečková P, Kopeček J. Antigen responsive hydrogels based on polymerizable antibody Fab′ fragment. Macromol Biosci. 2003;3(6):296–300.

[280] Miyata T, Asami N, Uragami T. Preparation of an antigen-sensitive hydrogel using antigen–antibody bindings. Macromolecules. 1999;32(6):2082–4.

[281] Miyata T, Asami N, Uragami T. A reversibly antigen-responsive hydrogel. Nature. 1999; 399(6738):766–9.

[282] Davaran S, Ghamkhari A, Alizadeh E, Massoumi B, Jaymand M. Novel dual stimuli-responsive ABC triblock copolymer: RAFT synthesis, "schizophrenic" micellization, and its performance as an anticancer drug delivery nanosystem. J Colloid Interface Sci. 2017;488:282–93.

[283] Ortiz de Solorzano I, Alejo T, Abad M, Bueno-Alejo C, Mendoza G, Andreu V, et al. Cleavable and thermo-responsive hybrid nanoparticles for on-demand drug delivery. J Colloid Interface Sci. 2019;533:171–81.

[284] Wang X, Zhang J, Wang Y, Wang C, Xiao J, Zhang Q, et al. Multi-responsive photothermal-chemotherapy with drug-loaded melanin-like nanoparticles for synergetic tumor ablation. Biomaterials. 2016;81:114–24.

[285] Duan Z, Cai H, Zhang H, Chen K, Li N, Xu Z, et al. PEGylated multistimuli-responsive dendritic prodrug-based nanoscale system for enhanced anticancer activity. ACS Appl Mater Interfaces. 2018;10(42):35770–83.

[286] Wang Z, Chen Z, Liu Z, Shi P, Dong K, Ju E, et al. A multi-stimuli responsive gold nanocage–hyaluronic platform for targeted photothermal and chemotherapy. Biomaterials. 2014; 35(36):9678–88.

[287] Hu W, Qiu L, Cheng L, Hu Q, Liu Y, Hu Z, et al. Redox and pH dual responsive poly (amidoamine) dendrimer-poly(ethylene glycol) conjugates for intracellular delivery of doxorubicin. Acta Biomater. 2016;36:241–53.

[288] Proksch E. pH in nature, humans and skin. J Dermatol. 2018;45(9):1044–52.

[289] Duan J-J, Zhang L-N. Robust and smart hydrogels based on natural polymers. Chinese J Polym Sci 2017;35(10):1165–80.

[290] Griffin M, Premakumar Y, Seifalian A, Butler PE, Szarko M. Biomechanical characterization of human soft tissues using indentation and tensile testing. J Vis Exp. 2016;2016(118):1–8.

[291] Chen F-M, Liu X, Polym P, Author S. Advancing biomaterials of human origin for tissue engineering HHS Public Access Author manuscript. Prog Polym Sci. 2016;53:86–168.

[292] Murniati R, Sutisna, Wibowo E, Rokhmat M, Iskandar F, Abdullah M. Natural rubber nanocomposite as human-tissue-mimicking materials for replacement cadaver in medical surgical practice. Procedia Eng. 2017;170(022):101–7.

[293] Vacanti JP, Langer R. Tissue engineering: The design and fabrication of living replacement devices for surgical reconstruction and transplantation. Lancet. 1999;354(SUPPL.1):32–4.

[294] Mellati A, Akhtari J. Injectable hydrogels: a review of injectability mechanisms and biomedical applications. Res Mol Med. 2019;6(4):1–19.

[295] Amini AA, Nair LS. Injectable hydrogels for bone and cartilage repair. Biomed Mater. 2012; 7(2): 24105.

[296] El-Sherbiny IM, Yacoub MH. Hydrogel scaffolds for tissue engineering: progress and challenges. Glob Cardiol Sci Pract. 2013;2013(3):38.

[297] Ahmed EM. Hydrogel: preparation, characterization, and applications: a review. J Adv Res. 2015;6(2):105–21.

[298] Chen G, Tang W, Wang X, Zhao X, Chen C, Zhu Z. Applications of hydrogels with special physical properties in biomedicine. Polymers. 2019;11(9):1–17.

[299] Spicer CD. Hydrogel scaffolds for tissue engineering: the importance of polymer choice. Polym Chem. 2020;11(2):184–219.

[300] Balakrishnan B, Banerjee R. Biopolymer-based hydrogels for cartilage tissue engineering. Chem Rev. 2011;111(8):4453–74.

[301] Fan M, Tan H. Biocompatible conjugation for biodegradable hydrogels as drug and cell scaffolds. Cogent Eng. 2020;7(1):1736407.

[302] Eslahi N, Abdorahim M, Simchi A. Smart polymeric hydrogels for cartilage tissue engineering: a review on the chemistry and biological functions. Biomacromolecules. 2016;17(11):3441–63.

[303] Tabata Y. Biomaterial technology for tissue engineering applications. J R Soc Interface. 2009;6(3):S311–S324.

[304] Kim SJ, Kim H Il, Park SJ, Kim IY, Lee SH, Lee TS, et al. Behavior in electric fields of smart hydrogels with potential application as bio-inspired actuators. Smart Mater Struct. 2005; 14(4):511–4.

[305] Klouda L, Mikos AG. Thermoresponsive hydrogels in biomedical applications. Eur J Pharm Biopharm. 2008;68(1):34–45.

[306] Oveissi F, Naficy S, Le TYL, Fletcher DF, Dehghani F. Polypeptide-affined interpenetrating hydrogels with tunable physical and mechanical properties. Biomater Sci. 2019;7(3):926–37.

[307] García-García JM, Liras M, Quijada-Garrido I, Gallardo A, París R. Swelling control in thermo-responsive hydrogels based on 2-(2-methoxyethoxy)ethyl methacrylate by crosslinking and copolymerization with N-isopropylacrylamide. Polym J. 2011;43(11):887–92.

[308] Anderson JM, Rodriguez A, Chang DT. Foreign body reaction to biomaterials. Semin Immunol. 2008;20(2):86–100.

[309] Hendrikse SIS, Spaans S, Meijer EW, Dankers PYW. Supramolecular platform stabilizing growth factors. Biomacromolecules. 2018;19(7):2610–7.

[310] Rodriguez MJ, Brown J, Giordano J, Lin SJ, Omenetto FG, Kaplan DL. Silk based bioinks for soft tissue reconstruction using 3-dimensional (3D) printing with in vitro and in vivo assessments. Biomaterials. 2017;117:105–15.

[311] Xiong R, Zhang Z, Chai W, Huang Y, Chrisey DB. Freeform drop-on-demand laser printing of 3D alginate and cellular constructs. Biofabrication. 2015;7(4):45011.

[312] Macková H, Plichta Z, Hlídková H, Sedláček O, Konefal R, Sadakbayeva Z, et al. Reductively degradable poly(2-hydroxyethyl methacrylate) hydrogels with oriented porosity for tissue engineering applications. ACS Appl Mater Interfaces. 2017;9(12):10544–53.

[313] Li L, Wang N, Jin X, Deng R, Nie S, Sun L, et al. Biodegradable and injectable in situ cross-linking chitosan-hyaluronic acid based hydrogels for postoperative adhesion prevention. Biomaterials. 2014;35(12):3903–17.

[314] Fisher SA, Baker AEG, Shoichet MS. Designing peptide and protein modified hydrogels: selecting the optimal conjugation strategy. J Am Chem Soc. 2017;139(22):7416–27.

[315] Blache U, Ehrbar M. Inspired by nature: hydrogels as versatile tools for vascular engineering. Adv Wound Care. 2018;7(7):232–46.

[316] Catoira MC, Fusaro L, Di Francesco D, Ramella M, Boccafoschi F. Overview of natural hydrogels for regenerative medicine applications. J Mater Sci Mater Med. 2019;30(10):1–10.

[317] Tan H, Marra KG. Injectable, Biodegradable hydrogels for tissue engineering applications. Materials. 2010;3(3):1746–67.

[318] Neves MI, Araújo M, Moroni L, Silva RMP da, Barrias CC. Glycosaminoglycan-inspired biomaterials for the development of bioactive hydrogel networks. Molecules. 2020; 25(4):978.

[319] Jin R, Moreira Teixeira LS, Dijkstra PJ, Karperien M, van Blitterswijk CA, Zhong ZY, et al. Injectable chitosan-based hydrogels for cartilage tissue engineering. Biomaterials. 2009; 30(13):2544–51.

[320] Stern R, Kogan G, Jedrzejas MJ, Šoltés L. The many ways to cleave hyaluronan. Biotechnol Adv. 2007;25(6):537–57.

[321] Bukhari SNA, Roswandi NL, Waqas M, Habib H, Hussain F, Khan S, et al. Hyaluronic acid, a promising skin rejuvenating biomedicine: A review of recent updates and pre-clinical and clinical investigations on cosmetic and nutricosmetic effects. Int J Biol Macromol. 2018;120:1682–95.

[322] Fakhari A, Berkland C. Applications and emerging trends of hyaluronic acid in tissue engineering, as a dermal filler and in osteoarthritis treatment. Acta Biomater. 2013; 9(7):7081–92.

[323] Tan W-H, Takeuchi S. Monodisperse alginate hydrogel microbeads for cell encapsulation. Adv Mater. 2007;19(18):2696–701.

[324] Lee KY, Mooney DJ. Alginate: Properties and biomedical applications. Prog Polym Sci. 2012;37(1):106–26.

[325] Schloss AC, Williams DM, Regan LJ. Protein-based hydrogels for tissue engineering. Adv Exp Med Biol. 2016;940:167–77.

[326] Castro APG, Laity P, Shariatzadeh M, Wittkowske C, Holland C, Lacroix D. Combined numerical and experimental biomechanical characterization of soft collagen hydrogel substrate. J Mater Sci Mater Med. 2016;27(4):1–9.

[327] Schuurman W, Levett PA, Pot MW, Weeren PR van, Dhert WJA, Hutmacher DW, et al. Gelatin-methacrylamide hydrogels as potential biomaterials for fabrication of tissue-engineered cartilage constructs. Macromol Biosci. 2013;13(5):551–61.

[328] Kapoor S, Kundu SC. Silk protein-based hydrogels: promising advanced materials for biomedical applications. Acta Biomater. 2016;31:17–32.

[329] Vepari C, Kaplan DL. Silk as a biomaterial. Prog Polym Sci. 2007;32(8–9):991–1007.

[330] Wang X, Liu C. Fibrin hydrogels for endothelialized liver tissue engineering with a predesigned vascular network. Polymers. 2018;10(10):1048.

[331] Silva R, Singh R, Sarker B, Papageorgiou DG, Juhasz JA, Roether JA, et al. Hybrid hydrogels based on keratin and alginate for tissue engineering. J Mater Chem B. 2014;2(33):5441–51.

[332] Xia L-W, Xie R, Ju X-J, Wang W, Chen Q, Chu L-Y. Nano-structured smart hydrogels with rapid response and high elasticity. Nat Commun. 2013;4(1):1–11.

[333] Stoop R. Smart biomaterials for tissue engineering of cartilage. Injury. 2008;39(1):77–87.

[334] Collett J, Crawford A, Hatton P V, Geoghegan M, Rimmer S. Thermally responsive polymeric hydrogel brushes: synthesis, physical properties and use for the culture of chondrocytes. J R Soc Interface. 2006;4(12):117–26.

[335] Patenaude M, Hoare T. Injectable, degradable thermoresponsive poly(N-isopropylacrylamide) hydrogels. ACS Macro Lett. 2012;1(3):409–13.

[336] Koffi AA, Agnely F, Ponchel G, Grossiord JL. Modulation of the rheological and mucoadhesive properties of thermosensitive poloxamer-based hydrogels intended for the rectal administration of quinine. Eur J Pharm Sci. 2006;27(4):328–35.

[337] Morgado PI, Lisboa PF, Ribeiro MP, Miguel SP, Simões PC, Correia IJ, et al. Poly(vinyl alcohol)/chitosan asymmetrical membranes: Highly controlled morphology toward the ideal wound dressing. J Memb Sci. 2014;469:262–71.

[338] Jiang S, Liu S, Feng W. PVA hydrogel properties for biomedical application. J Mech Behav Biomed Mater. 2011;4(7):1228–33.

[339] Zhu J. Bioactive modification of poly(ethylene glycol) hydrogels for tissue engineering. Biomaterials. 2010;31(17):4639–56.

[340] Lin C-C, Anseth KS. PEG Hydrogels for the controlled release of biomolecules in regenerative medicine. Pharm Res. 2008;26(3):631–43.

[341] Reichert JC, Heymer A, Berner A, Eulert J, Nöth U. Fabrication of polycaprolactone collagen hydrogel constructs seeded with mesenchymal stem cells for bone regeneration. Biomed Mater. 2009;4(6):065001.

[342] Hernandez I, Kumar A, Joddar B. A bioactive hydrogel and 3D printed polycaprolactone system for bone tissue engineering. Gels. 2017;3(3):26.

[343] Basu A, Kunduru KR, Doppalapudi S, Domb AJ, Khan W. Poly(lactic acid) based hydrogels. Adv Drug Deliv Rev. 2016;107:192–205.

[344] Shi K, Wang Y-L, Qu Y, Liao J-F, Chu B-Y, Zhang H-P, et al. Synthesis, characterization and application of reversible PDLLA-PEG-PDLLA copolymer thermogels in vitro and in vivo. Sci Rep. 2016;6(1):1–15.

[345] Lim HL, Hwang Y, Kar M, Varghese S. Smart hydrogels as functional biomimetic systems. Biomater Sci. 2014;2(5):603–18.

[346] Prabaharan M, Mano JF. Stimuli-responsive hydrogels based on polysaccharides incorporated with thermo-responsive polymers as novel biomaterials. Macromol Biosci. 2006;6(12):991–1008.

[347] Choi A, Seo KD, Yoon H, Han SJ, Kim DS. Bulk poly(N-isopropylacrylamide) (PNIPAAm) thermoresponsive cell culture platform: Toward a new horizon in cell sheet engineering. Biomater Sci. 2019;7(6):2277–87.

[348] Sá-Lima H, Tuzlakoglu K, Mano JF, Reis RL. Thermoresponsive poly(N-isopropylacrylamide)-g-methylcellulose hydrogel as a three-dimensional extracellular matrix for cartilage-engineered applications. J Biomed Mater Res A. 2011;98 A(4):596–603.

[349] Avais M, Chattopadhyay S. Waterborne pH responsive hydrogels: Synthesis, characterization and selective pH responsive behavior around physiological pH. Polymer. 2019;180:121701.

[350] Rizwan M, Yahya R, Hassan A, Yar M, Azzahari AD, Selvanathan V, et al. pH sensitive hydrogels in drug delivery: brief history, properties, swelling, and release mechanism, material selection and applications. Polymers. 2017;9(4):137.

[351] Adams DJ, Adams S, Atkins D, Butler MF, Furzeland S. Impact of mechanism of formation on encapsulation in block copolymer vesicles. J Control Release. 2008;128(2):165–70.

[352] Strehin I, Nahas Z, Arora K, Nguyen T, Elisseeff J. A versatile pH sensitive chondroitin sulfate-PEG tissue adhesive and hydrogel. Biomaterials. 2010;31(10):2788–97.

[353] Kocak FZ, Talari ACS, Yar M, Rehman IU. In-situ forming pH and thermosensitive injectable hydrogels to stimulate angiogenesis: Potential candidates for fast bone regeneration applications. Int J Mol Sci. 2020;21(5):1633.

[354] Li L, Scheiger JM, Levkin PA. Design and applications of photoresponsive hydrogels. Adv Mater. 2019;31(26):1807333.

[355] Tomatsu I, Peng K, Kros A. Photoresponsive hydrogels for biomedical applications. Adv Drug Deliv Rev. 2011;63(14–15):1257–66.

[356] Anseth KS, Metters AT, Bryant SJ, Martens PJ, Elisseeff JH, Bowman CN. In situ forming degradable networks and their application in tissue engineering and drug delivery. J Control Release. 2002;78(1–3):199–209.

[357] Levett PA, Melchels FPW, Schrobback K, Hutmacher DW, Malda J, Klein TJ. A biomimetic extracellular matrix for cartilage tissue engineering centered on photocurable gelatin, hyaluronic acid and chondroitin sulfate. Acta Biomater. 2014;10(1):214–23.

[358] Li L, Ge J, Guo B, Ma PX. In situ forming biodegradable electroactive hydrogels. Polym Chem. 2014;5(8):2880–90.

[359] Rahimi N, Swennen G, Verbruggen S, Scibiorek M, Molin DG, Post MJ. Short stimulation of electro-responsive PAA/fibrin hydrogel induces collagen production. Tissue Eng Part C Methods. 2014;20(9):703–13.

[360] Bagheri B, Zarrintaj P, Surwase SS, Baheiraei N, Saeb MR, Mozafari M, et al. Self-gelling electroactive hydrogels based on chitosan–aniline oligomers/agarose for neural tissue engineering with on-demand drug release. Colloids Surfaces B. 2019;184:110549.

[361] Zhang Y, Yu J, Bomba HN, Zhu Y, Gu Z. Mechanical force-triggered drug delivery. Chem Rev. 2016;116(19):12536–63.

[362] Discher DE, Janmey P, Wang YL. Tissue cells feel and respond to the stiffness of their substrate. Science. 2005;310(5751):1139–43.

[363] O'Conor CJ, Case N, Guilak F. Mechanical regulation of chondrogenesis. Stem Cell Res Ther. 2013;4(4):61.

[364] Lim HL, Chuang JC, Tran T, Aung A, Arya G, Varghese S. Dynamic electromechanical hydrogel matrices for stem cell culture. Adv Funct Mater. 2011;21(1):55–63.

[365] Lin S, Liu J, Liu X, Zhao X. Muscle-like fatigue-resistant hydrogels by mechanical training. Proc Natl Acad Sci U S A. 2019;116(21):10244–9.

[366] Chaturvedi K, Ganguly K, Nadagouda MN, Aminabhavi TM. Polymeric hydrogels for oral insulin delivery. J Control Release. 2013;165(2):129–38.

[367] Maitz MF, Freudenberg U, Tsurkan M V., Fischer M, Beyrich T, Werner C. Bio-responsive polymer hydrogels homeostatically regulate blood coagulation. Nat Commun. 2013;4:1–7.

[368] Skaalure SC, Chu S, Bryant SJ. An enzyme-sensitive PEG hydrogel based on aggrecan catabolism for cartilage tissue engineering. Adv Healthc Mater. 2015;4(3):420–31.

[369] Knipe JM, Peppas NA. Multi-responsive hydrogels for drug delivery and tissue engineering applications. Regen Biomater. 2014;1(1):57–65.

[370] Sengupta D, Waldman SD, Li S. From in vitro to in situ tissue engineering. Ann Biomed Eng. 2014;42(7):1537–45.

[371] Huang S, Yang Y, Yang Q, Zhao Q, Ye X. Engineered circulatory scaffolds for building cardiac tissue. J Thorac Dis. 2018;10(I):S2312–28.

[372] Adelöw CAM, Frey P. Synthetic hydrogel matrices for guided bladder tissue regeneration. Methods Mol Med. 2007;140:125–40.

[373] Meng K, Yao C, Ma Q, Xue Z, Du Y, Liu W, et al. A reversibly responsive fluorochromic hydrogel based on lanthanide–mannose complex. Adv Sci. 2019;6(10):1802112.

[374] Sackett SD, Tremmel DM, Ma F, Feeney AK, Maguire RM, Brown ME, et al. Extracellular matrix scaffold and hydrogel derived from decellularized and delipidized human pancreas. Sci Rep. 2018;8(1):1–16.

[375] Wilkinson AC, Ishida R, Kikuchi M, Sudo K, Morita M, Crisostomo RV, et al. Long-term ex vivo haematopoietic-stem-cell expansion allows nonconditioned transplantation. Nature. 2019;571(7763):117–21.

[376] Babensee JE, Anderson JM, McIntire L V., Mikos AG. Host response to tissue engineered devices. Adv Drug Deliv Rev. 1998;33(1–2):111–39.

[377] Sengupta D, Gilbert PM, Johnson KJ, Blau HM, Heilshorn SC. Protein-engineered biomaterials to generate human skeletal muscle mimics. Adv Healthc Mater. 2012;1(6):785–9.

[378] Erickson CB, Newsom JP, Fletcher NA, Feuer ZM, Yu Y, Rodriguez-Fontan F, et al. In vivo degradation rate of alginate–chitosan hydrogels influences tissue repair following physeal injury. J Biomed Mater Res B. 2020;108(6):2484–94.

[379] Yahia Lh. History and applications of hydrogels. J Biomed Sci. 2015;4(2):1–23.

[380] Sivashanmugam A, Arun Kumar R, Vishnu Priya M, Nair S V., Jayakumar R. An overview of injectable polymeric hydrogels for tissue engineering. Eur Polym J. 2015;72:543–65.

[381] Talebian S, Mehrali M, Taebnia N, Pennisi CP, Kadumudi FB, Foroughi J, et al. Self-healing hydrogels: the next paradigm shift in tissue engineering? Adv Sci. 2019;6(16):1801664.

[382] Liu Y, Lim J, Teoh SH. Review: development of clinically relevant scaffolds for vascularised bone tissue engineering. Biotechnol Adv. 2013;31(5):688–705.

[383] Huebsch N, Lippens E, Lee K, Mehta M, Koshy ST, Darnell MC, et al. Matrix elasticity of void-forming hydrogels controls transplanted-stem-cell-mediated bone formation. Nat Mater. 2015;14(12):1269–77.

[384] Shi L, Wang F, Zhu W, Xu Z, Fuchs S, Hilborn J, et al. Self-healing silk fibroin-based hydrogel for bone regeneration: dynamic metal-ligand self-assembly approach. Adv Funct Mater. 2017;27(37):1–14.

[385] Zhang Y, Chen M, Tian J, Gu P, Cao H, Fan X, et al. In situ bone regeneration enabled by a biodegradable hybrid double-network hydrogel. Biomater Sci. 2019;7(8):3266–76.

[386] Gačanin J, Hedrich J, Sieste S, Glaßer G, Lieberwirth I, Schilling C, et al. Autonomous ultrafast self-healing hydrogels by pH-responsive functional nanofiber gelators as cell matrices. Adv Mater. 2019;31(2):1805044.

[387] Sridhar B V., Brock JL, Silver JS, Leight JL, Randolph MA, Anseth KS. Development of a cellularly degradable PEG hydrogel to promote articular cartilage extracellular matrix deposition. Adv Healthc Mater. 2015;4(5):702–13.

[388] Biondi M, Borzacchiello A, Mayol L, Ambrosio L. Nanoparticle-integrated hydrogels as multifunctional composite materials for biomedical applications. Gels. 2015;1(2):162–78.

[389] Wang Z, Roberge C, Wan Y, Dao LH, Guidoin R, Zhang Z. A biodegradable electrical bioconductor made of polypyrrole nanoparticle/poly(D,L-lactide) composite: a preliminary in vitro biostability study. J Biomed Mater Res A. 2003;66(4):738–46.

[390] Kašpárková V, Humpolíček P, Stejskal J, Capáková Z, Bober P, Skopalová K, et al. Exploring the critical factors limiting polyaniline biocompatibility. Polymers. 2019;11(2):362.

[391] He H, Zhang L, Guan X, Cheng H, Liu X, Yu S, et al. Biocompatible conductive polymers with high conductivity and high stretchabiliy. ACS Appl Mater Interfaces. 2019;11(29):26185–93.

[392] Zarrintaj P, Moghaddam AS, Manouchehri S, Atoufi Z, Amiri A, Amirkhani MA, et al. Can regenerative medicine and nanotechnology combine to heal wounds? the search for the ideal wound dressing. Nanomedicine. 2017;12(19):2403–22.

[393] Jeong KH, Park D, Lee YC. Polymer-based hydrogel scaffolds for skin tissue engineering applications: a mini-review. J Polym Res. 2017;24:112.

[394] Kamoun EA, Kenawy ERS, Chen X. A review on polymeric hydrogel membranes for wound dressing applications: PVA-based hydrogel dressings. J Adv Res. 2017;8(3):217–33.

[395] Zhao X, Wu H, Guo B, Dong R, Qiu Y, Ma PX. Antibacterial anti-oxidant electroactive injectable hydrogel as self-healing wound dressing with hemostasis and adhesiveness for cutaneous wound healing. Biomaterials. 2017;122:34–47.

[396] Gong CY, Wu QJ, Wang YJ, Zhang DD, Luo F, Zhao X, et al. A biodegradable hydrogel system containing curcumin encapsulated in micelles for cutaneous wound healing. Biomaterials. 2013;34(27):6377–87.

[397] Han L, Zhang Y, Lu X, Wang K, Wang Z, Zhang H. Polydopamine nanoparticles modulating stimuli-responsive PNIPAM Hydrogels with Cell/Tissue Adhesiveness. ACS Appl Mater Interfaces. 2016;8(42):29088–100.

[398] Hasan A, Khattab A, Islam MA, Hweij KA, Zeitouny J, Waters R, et al. Injectable hydrogels for cardiac tissue repair after myocardial infarction. Adv Sci. 2015;2(11):1–18.

[399] Koudstaal S, Bastings MMC, Feyen DAM, Waring CD, Van Slochteren FJ, Dankers PYW, et al. Sustained delivery of insulin-like growth factor-1/hepatocyte growth factor stimulates endogenous cardiac repair in the chronic infarcted pig heart. J Cardiovasc Transl Res. 2014; 7(2):232–41.

[400] Dong R, Zhao X, Guo B, Ma PX. Self-healing conductive injectable hydrogels with antibacterial activity as cell delivery carrier for cardiac cell therapy. ACS Appl Mater Interfaces. 2016;8(27):17138–50.

[401] Pakulska MM, Ballios BG, Shoichet MS. Injectable hydrogels for central nervous system therapy. Biomed Mater. 2012;7(2):24101.

[402] Qin H, Zhang T, Li N, Cong HP, Yu SH. Anisotropic and self-healing hydrogels with multi-responsive actuating capability. Nat Commun. 2019;10(1):1–11.

[403] Jacob RS, Ghosh D, Singh PK, Basu SK, Jha NN, Das S, et al. Self healing hydrogels composed of amyloid nano fibrils for cell culture and stem cell differentiation. Biomaterials. 2015;54:97–105.

[404] Tseng TC, Tao L, Hsieh FY, Wei Y, Chiu IM, Hsu SH. An injectable, self-healing hydrogel to repair the central nervous system. Adv Mater. 2015;27(23):3518–24.

[405] Wei Z, Zhao J, Chen YM, Zhang P, Zhang Q. Self-healing polysaccharide-based hydrogels as injectable carriers for neural stem cells. Sci Rep. 2016;6:1–12.

[406] Hou C, Duan Y, Zhang Q, Wang H, Li Y. Bio-applicable and electroactive near-infrared laser-triggered self-healing hydrogels based on graphene networks. J Mater Chem. 2012; 22(30):14991–6.

[407] Lee ES, Shin JM, Son S, Ko H, Um W, Song SH, et al. Recent advances in polymeric nanomedicines for cancer immunotherapy. Adv Healthc Mater. 2019;8(4):1–44.

[408] Carelle N, Piotto E, Bellanger A, Germanaud J, Thuillier A, Khayat D. Changing patient perceptions of the side effects of cancer chemotherapy. Cancer. 2002;95(1):155–63.

[409] Liu M, Guo F. Recent updates on cancer immunotherapy. Precis Clin Med. 2018;1(2):65–74.

[410] Brookmeyer R, Johnson E, Ziegler-Graham K, Arrighi HM. Forecasting the global burden of Alzheimer's disease. Alzheimer's Dement. 2007;3(3):186–91.

[411] Gómez-Mascaraque LG, Palao-Suay R, Vázquez B. The use of smart polymers in medical devices for minimally invasive surgery, diagnosis and other applications. Aguilar MR, San Román J, editor. Smart Polymers and their Applications. Woodhead Publishing; 2014. p. 359–407. Ch. 12, pp 359-407.

[412] Maitland DJ, Metzger MF, Schumann D, Lee A, Wilson TS. Photothermal properties of shape memory polymer micro-actuators for treating stroke. Lasers Surg Med. 2002;30(1):1–11.

[413] Sokolowski W, Metcalfe A, Hayashi S, Yahia L, Raymond J. Medical applications of shape memory polymers. Biomed Mater. 2007;2(1):S23.

[414] Yakacki CM, Shandas R, Lanning C, Rech B, Eckstein A, Gall K. Unconstrained recovery characterization of shape-memory polymer networks for cardiovascular applications. Biomaterials. 2007;28(14):2255–63.

[415] Xie Y, Lei D, Wang S, Liu Z, Sun L, Zhang J, et al. A biocompatible, biodegradable, and functionalizable copolyester and its application in water-responsive shape memory scaffold. ACS Biomater Sci Eng. 2019;5(4):1668–76.

[416] Lv H, Tang D, Sun Z, Gao J, Yang X, Jia S, et al. Electrospun PCL-based polyurethane/HA microfibers as drug carrier of dexamethasone with enhanced biodegradability and shape memory performances. Colloid Polym Sci. 2020;298(1):103–11.

[417] Kratz K, Voigt U, Lendlein A. Temperature-memory effect of copolyesterurethanes and their application potential in minimally invasive medical technologies. Adv Funct Mater. 2012; 22(14):3057–65.

[418] Jung F, Braune S. Thrombogenicity and hemocompatibility of biomaterials. Biointerphases. 2016;11(2):029601.

[419] Peterson GI, Dobrynin A V., Becker ML. Biodegradable shape memory polymers in medicine. Adv Healthc Mater. 2017;6(21):1–16.

[420] Kashif M, Yun BM, Lee KS, Chang YW. Biodegradable shape-memory poly(ε-caprolactone)/ polyhedral oligomeric silsequioxane nanocomposites: sustained drug release and hydrolytic degradation. Mater Lett. 2016;166:125–8.

[421] Guo B, Chen Y, Lei Y, Zhang L, Zhou WY, Rabie ABM, et al. Biobased poly(propylene sebacate) as shape memory polymer with tunable switching temperature for potential biomedical applications. Biomacromolecules. 2011;12(4):1312–21.

[422] Wu J, Yuan C, Ding Z, Isakov M, Mao Y, Wang T, et al. Multi-shape active composites by 3D printing of digital shape memory polymers. Sci Rep. 2016;6:1–11.

[423] Boire TC, Gupta MK, Zachman AL, Lee SH, Balikov DA, Kim K, et al. Reprint of: pendant allyl crosslinking as a tunable shape memory actuator for vascular applications. Acta Biomater. 2016;34:73–83.

[424] Hearon K, Wierzbicki MA, Nash LD, Landsman TL, Laramy C, Lonnecker AT, et al. A processable shape memory polymer system for biomedical applications. Adv Healthc Mater. 2015;4(9):1386–98.

[425] Lendlein A, Langer R. Biodegradable, elastic shape-memory polymers for potential biomedical applications. Science. 2002;296(5573):1673–6.

[426] Dangas GD, Claessen BE, Caixeta A, Sanidas EA, Mintz GS, Mehran R. In-stent restenosis in the drug-eluting stent era. J Am Coll Cardiol. 2010;56(23):1897–907.

[427] Xue L, Dai S, Li Z. Synthesis and characterization of elastic star shape-memory polymers as self-expandable drug-eluting stents. J Mater Chem. 2012;22(15):7403–11.

[428] Ajili SH, Ebrahimi NG, Soleimani M. Polyurethane/polycaprolactane blend with shape memory effect as a proposed material for cardiovascular implants. Acta Biomater. 2009; 5(5):1519–30.

[429] Chen MC, Tsai HW, Chang Y, Lai WY, Mi FL, Liu CT, et al. Rapidly self-expandable polymeric stents with a shape-memory property. Biomacromolecules. 2007;8(9):2774–80.

[430] Soares JS, Moore JE. Biomechanical challenges to polymeric biodegradable stents. Ann Biomed Eng. 2016;44(2):560–79.

[431] Wang J, Luo J, Kunkel R, Saha M, Bohnstedt BN, Lee CH, et al. Development of shape memory polymer nanocomposite foam for treatment of intracranial aneurysms. Mater Lett. 2019;250:38–41.

[432] Herting SM, Ding Y, Boyle AJ, Dai D, Nash LD, Asnafi S, et al. In vivo comparison of shape memory polymer foam-coated and bare metal coils for aneurysm occlusion in the rabbit elastase model. J Biomed Mater Res B. 2019;107(8):2466–75.

[433] Subash A, Kandasubramanian B. 4D printing of shape memory polymers. Eur Polym J. 2020;134:109771.

[434] Castro NJ, Meinert C, Levett P, Hutmacher DW. Current developments in multifunctional smart materials for 3D/4D bioprinting. Curr Opin Biomed Eng. 2017;2:67–75.

[435] Chan V, Zorlutuna P, Jeong JH, Kong H, Bashir R. Three-dimensional photopatterning of hydrogels using stereolithography for long-term cell encapsulation. Lab Chip. 2010; 10(16):2062–70.

[436] Guo F, Mao Z, Chen Y, Xie Z, Lata JP, Li P, et al. Three-dimensional manipulation of single cells using surface acoustic waves. Proc Natl Acad Sci USA. 2016;113(6):1522–7.

[437] Matai I, Kaur G, Seyedsalehi A, McClinton A, Laurencin CT. Progress in 3D bioprinting technology for tissue/organ regenerative engineering. Biomaterials. 2020;226:119536.

[438] Lee V, Singh G, Trasatti JP, Bjornsson C, Xu X, Tran TN, et al. Design and fabrication of human skin by three-dimensional bioprinting. Tissue Eng Part C Methods. 2014;20(6):473–84.

[439] Das S, Pati F, Choi YJ, Rijal G, Shim JH, Kim SW, et al. Bioprintable, cell-laden silk fibroin-gelatin hydrogel supporting multilineage differentiation of stem cells for fabrication of three-dimensional tissue constructs. Acta Biomater. 2015;11(1):233–46.

[440] Gao G, Schilling AF, Yonezawa T, Wang J, Dai G, Cui X. Bioactive nanoparticles stimulate bone tissue formation in bioprinted three-dimensional scaffold and human mesenchymal stem cells. Biotechnol J. 2014;9(10):1304–11.

[441] Kapoor S, Kundu SC. Silk protein-based hydrogels: Promising advanced materials for biomedical applications. Acta Biomater. 2016;31:17–32.

[442] Law N, Doney B, Glover H, Qin Y, Aman ZM, Sercombe TB, et al. Characterisation of hyaluronic acid methylcellulose hydrogels for 3D bioprinting. J Mech Behav Biomed Mater. 2018;77:389–99.

[443] Freeman S, Ramos R, Alexis Chando P, Zhou L, Reeser K, Jin S, et al. A bioink blend for rotary 3D bioprinting tissue engineered small-diameter vascular constructs. Acta Biomater. 2019;95:152–64.

[444] Chen Y, Zhang J, Liu X, Wang S, Tao J, Huang Y, et al. Noninvasive in vivo 3D bioprinting. Sci Adv. 2020;6(23):1–11.

[445] Xu F, Celli J, Rizvi I, Moon S, Hasan T, Demirci U. A three-dimensional in vitro ovarian cancer coculture model using a high-throughput cell patterning platform. Biotechnol J. 2011; 6(2):204–12.

[446] Mao S, Pang Y, Liu T, Shao Y, He J, Yang H, et al. Bioprinting of in vitro tumor models for personalized cancer treatment: a review. Biofabrication. 2020;12(4):42001.

[447] Webster A, Greenman J, Haswell SJ. Development of microfluidic devices for biomedical and clinical application. J Chem Technol Biotechnol. 2011;86(1):10–7.

[448] Cimrák I, Gusenbauer M, Schrefl T. Modelling and simulation of processes in microfluidic devices for biomedical applications. Comput Math with Appl. 2012;64(3):278–88.

[449] Tiwari SK, Bhat S, Mahato KK. Design and fabrication of low-cost microfluidic channel for biomedical application. Sci Rep. 2020;10(1):1–14.

[450] Hilber W. Stimulus-active polymer actuators for next-generation microfluidic devices. Appl Phys A. 2016;122(8):1–39.

[451] Kieviet BD, Schön PM, Vancso GJ. Stimulus-responsive polymers and other functional polymer surfaces as components in glass microfluidic channels. Lab Chip. 2014;14(21): 4159–70.

[452] Jiang C, Takehara H, Uto K, Ebara M, Aoyagi T, Ichiki T. Evaluation of microvalves developed for point-of-care testing devices using shape-memory polymers. J Photopolym Sci Technol. 2013;26(5):581–5.

[453] Kim J, Baek JY, Lee K, Park YD, Sun K, Lee TS, et al. Photopolymerized check valve and its integration into a pneumatic pumping system for biocompatible sample delivery. Lab Chip. 2006;6(8):1091–4.

[454] Obst F, Beck A, Bishayee C, Mehner PJ, Richter A, Voit B, et al. Hydrogel microvalves as control elements for parallelized enzymatic cascade reactions in microfluidics. Micromachines. 2020;11(2):167.

[455] Agarwal AK, Sridharamurthy SS, Beebe DJ, Jiang H. Programmable autonomous micromixers and micropumps. J Microelectromechanical Syst. 2005;14(6):1409–21.

[456] Hara Y, Yoshida R. Self-oscillating polymer fueled by organic acid. J Phys Chem B. 2008; 112(29):8427–9.

[457] Kim H, Kim K, Lee SJ. Compact and thermosensitive nature-inspired micropump. Sci Rep. 2016;6:1–10.

[458] Seo J, Wang C, Chang S, Park J, Kim W. A hydrogel-driven microfluidic suction pump with a high flow rate. Lab Chip. 2019;19(10):1790–6.

[459] Dickens E, Ahmed S. Principles of cancer treatment by chemotherapy. Surg (United Kingdom). 2018;36(3):134–8.

[460] Szewczuk MR. Advancements in Polymer Science: 'Smart' drug delivery systems for the treatment of cancer. MOJ Polym Sci. 2017;1(3):113–8.

[461] Thambi T, Park JH, Lee DS. Stimuli-responsive polymersomes for cancer therapy. Biomater Sci. 2016;4(1):55–69.

[462] Morales-Cruz M, Delgado Y, Castillo B, Figueroa CM, Molina AM, Torres A, et al. Smart targeting to improve cancer therapeutics. Drug Des Devel Ther. 2019;13:3753–72.

[463] Nishiyama N, Bae Y, Miyata K, Fukushima S, Kataoka K. Smart polymeric micelles for gene and drug delivery. Drug Discov Today Technol. 2005;2(1):21–6.

[464] Zhou L, Wang H, Li Y. Stimuli-responsive nanomedicines for overcoming cancer multidrug resistance. Theranostics. 2018;8(4):1059–74.

[465] Rao NV, Ko H, Lee J, Park JH. Recent progress and advances in stimuli-responsive polymers for cancer therapy. Front Bioeng Biotechnol. 2018;6:110.

[466] Chang G, Li C, Lu W, Ding J. N-Boc-Histidine-capped PLGA-PEG-PLGA as a smart polymer for drug delivery sensitive to tumor extracellular pH. Macromol Biosci. 2010;10(10):1248–56.

[467] Kamada H, Tsutsumi Y, Yoshioka Y, Yamamoto Y, Kodaira H, Tsunoda SI, et al. Design of a pH-sensitive polymeric carrier for drug release and its application in cancer therapy. Clin Cancer Res. 2004;10(7):2545–50.

[468] Salgarella AR, Zahoranová A, Šrámková P, Majerčíková M, Pavlova E, Luxenhofer R, et al. Investigation of drug release modulation from poly(2-oxazoline) micelles through ultrasound. Sci Rep. 2018;8(1):1–13.

[469] Ma P, Mumper RJ. Paclitaxel nano-delivery systems: a comprehensive review. J Nanomedicine Nanotechnol. 2013;4(2):1000164.

[470] Meng F, Cheng R, Deng C, Zhong Z. Intracellular drug release nanosystems. Mater Today. 2012;15(10):436–42.

[471] Zhai S, Hu X, Hu Y, Wu B, Xing D. Visible light-induced crosslinking and physiological stabilization of diselenide-rich nanoparticles for redox-responsive drug release and combination chemotherapy. Biomaterials. 2017;121:41–54.

[472] Han HS, Thambi T, Choi KY, Son S, Ko H, Lee MC, et al. Bioreducible shell-cross-linked hyaluronic acid nanoparticles for tumor-targeted drug delivery. Biomacromolecules. 2015; 16(2):447–56.

[473] Parker N, Turk MJ, Westrick E, Lewis JD, Low PS, Leamon CP. Folate receptor expression in carcinomas and normal tissues determined by a quantitative radioligand binding assay. Anal Biochem. 2005;338(2):284–93.

[474] Li X, Sambi M, Decarlo A, Burov S V., Akasov R, Markvicheva E, et al. Functionalized folic acid-conjugated amphiphilic alternating copolymer actively targets 3D multicellular tumour spheroids and delivers the hydrophobic drug to the inner core. Nanomaterials. 2018;8(8): 1–21.

[475] Giordano A, Tommonaro G. Curcumin and cancer. Nutrients. 2019;11(10):2376.

[476] Mahalunkar S, Yadav AS, Gorain M, Pawar V, Braathen R, Weiss S, et al. Functional design of pH-responsive folate-targeted polymer-coated gold nanoparticles for drug delivery and in vivo therapy in breast cancer. Int J Nanomedicine. 2019;14:8285–302.

[477] Kamphorst AO, Guermonprez P, Dudziak D, Nussenzweig MC. Route of antigen uptake differentially impacts presentation by dendritic cells and activated monocytes. J Immunol. 2010;185(6):3426–35.

[478] Qiu F, Becker KW, Knight FC, Baljon JJ, Sevimli S, Shae D, et al. Poly(propylacrylic acid)-peptide nanoplexes as a platform for enhancing the immunogenicity of neoantigen cancer vaccines. Biomaterials. 2018;182:82–91.

[479] Kim YC, Park JH, Prausnitz MR. Microneedles for drug and vaccine delivery. Adv Drug Deliv Rev. 2012;64(14):1547–68.

[480] De Geest BG, Willart MA, Hammad H, Lambrecht BN, Pollard C, Bogaert P, et al. Polymeric multilayer capsule-mediated vaccination induces protective immunity against cancer and viral infection. ACS Nano. 2012;6(3):2136–49.

[481] Duong HTT, Yin Y, Thambi T, Nguyen TL, Giang Phan VH, Lee MS, et al. Smart vaccine delivery based on microneedle arrays decorated with ultra-pH-responsive copolymers for cancer immunotherapy. Biomaterials. 2018;185:13–24.

[482] Dugger BN, Dickson DW. Pathology of neurodegenerative diseases. Cold Spring Harb Perspect Biol. 2017;9(7):1–22.

[483] Rad Mansoor K, Rad Sima K, Rad Soheila K. Advancement of polymer–based nanoparticles as smart drug delivery systems in neurodegenerative medicine. J Nanomedicine Res. 2019; 8(1):8–11.

[484] Gaudin A, Andrieux K, Couvreur P. Nanomedicines and stroke: toward translational research. J Drug Deliv Sci Technol. 2015;30:278–99.

[485] Upadhyay RK. Drug delivery systems, CNS protection, and the blood brain barrier. Biomed Res Int. 2014;2014:869269.

[486] Harilal S, Jose J, Parambi DGT, Kumar R, Mathew GE, Uddin MS, et al. Advancements in nanotherapeutics for Alzheimer's disease: current perspectives. J Pharm Pharmacol. 2019; 71(9):1370–83.

[487] Kulkarni P V., Roney CA, Antich PP, Bonte FJ, Raghu A V., Aminabhavi TM. Quinoline-n-butylcyanoacrylate-based nanoparticles for brain targeting for the diagnosis of Alzheimer's disease. Wiley Interdiscip Rev Nanomed Nanobiotechnol. 2010;2(1):35–47.

[488] Wilson B, Samanta MK, Muthu MS, Vinothapooshan G. Design and evaluation of chitosan nanoparticles as novel drug carrier for the delivery of rivastigmine to treat Alzheimer's disease. Ther Deliv. 2011;2(5):599–609.

[489] Lipp L, Sharma D, Banerjee A, Singh J. In Vitro and in vivo optimization of phase sensitive smart polymer for controlled delivery of rivastigmine for treatment of Alzheimer's disease. Pharm Res. 2020;37:34.

Index

https://doi.org/10.1515/9781501522468-007